Super Structures

Engineers today are building some of the most complicated structures ever seen. In this book you can read the exciting story of how many of them are built. Page after page of cutaway illustrations and diagrams explain how huge machines such as the 'mole' work, and the problems involved in creating structures as different as the Empire State Building and Olympic Stadium in Munich. And for budding technologists, the science of building comes to life with simple projects and experiments.

The Viking Press New York

Contents

Information stars

A star★ in an introduction indicates a know-how box. These boxes will explain how things work. Look for a star in a box like the one shown here.

Red boxes

Red boxes point out projects to make and quick experiments to try. There are simple instructions and diagrams to follow for each one. All you will need are a few easy-to-get materials.

Start to finish

Removing the soil from the site.
tracked excavator
articulated wheeled loader
excavator
Digging the foundation.
site truck
fixed site crane
ready-mixed-concrete truck
air compressor

People and machines

The super structures of today take a vast team to build them. Three of the most important members of this team are shown below.

The architect makes sure the building is strong enough.

The engineer makes sure the whole job goes smoothly.

The workman does all the jobs machines cannot do.

If you could walk onto a building site you would probably be astonished at all the different kinds of machines in use. But you can be sure that every machine is needed. When the architect has completed his designs, the site engineer examines them. He has to decide from these plans just what machinery he will need to build the designs the architect has drawn.

On site

The earth-moving machines are first to arrive—bulldozers, excavators and scrapers. Their job is to shift great mounds of earth from one part of a site to another, or take it away completely. When the ground has been prepared, the builder sets up a concrete mixer. If the site is small and he needs a lot of concrete quickly, it may have to be delivered ready-mixed in big drums on the backs of trucks. As the structure grows, cranes will be brought in to lift material up to the top.

Workmen are a vital part of the building team. Without them, no building could be erected. They perform countless tasks that machines simply cannot carry out.

Below: The mobile crane can move quickly to a site.

Below: A scraper digs, loads and takes the soil away.

Below: For small jobs an excavator arm on a tractor is often used.

Above: The tracked excavator can move over any ground. **Above: A tracked loader tips out some soil.**

Below: Inside the towers, sand, stone and cement are mixed together; trucks take this mixture to the site.

Methods and materials

Bricks and masonry

English pattern

Flemish pattern

The simple method of building a house with bricks or blocks of stone has been known for centuries. Bricks are blocks of clay that have been baked hard in a hot oven. The bricks in a wall are held together with mortar. Different patterns of brick are sometimes made to give added strength to a wall.

Steel

Steel beams or girders can be made in many shapes and sizes and fitted together like a construction set. Girders are either bolted or welded. A welded joint is made by heating two pieces of metal together with a welding rod. The rod melts first and flows around the joint area. When this cools, a strong bond is formed.

Reinforced concrete

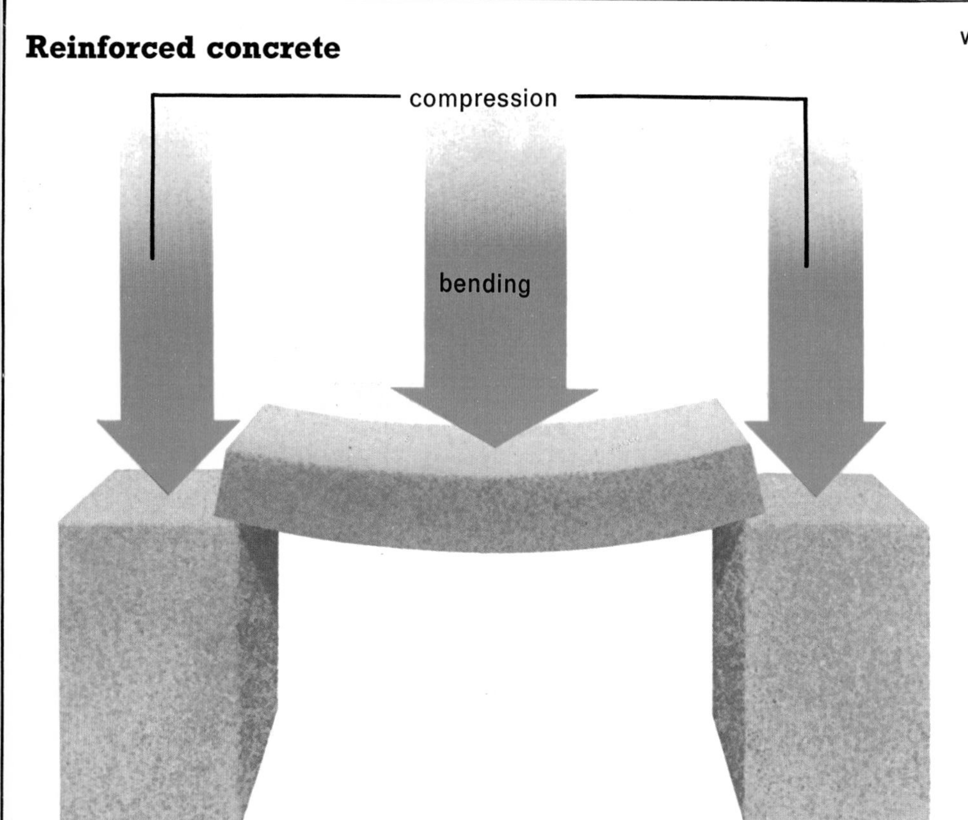

wooden form

finished reinforced concrete beam

Today, concrete is the most widely used building material. Concrete is sand, stone and cement mixed with water. The cement and water combine to make a strong glue that sticks all the stones and sand together as one. Concrete is very strong when it is compressed, but if bent, it breaks more easily. To withstand the force of being bent, steel bars are added to concrete. This mixture is called reinforced concrete.

When it is first mixed, concrete is wet and runny. It sets hard in about a day. To make a shape from concrete, it is poured into a wooden or steel mold called a *form*. The reinforcing steel bars are put in their place first and the concrete poured around them. This is known as *casting* concrete.

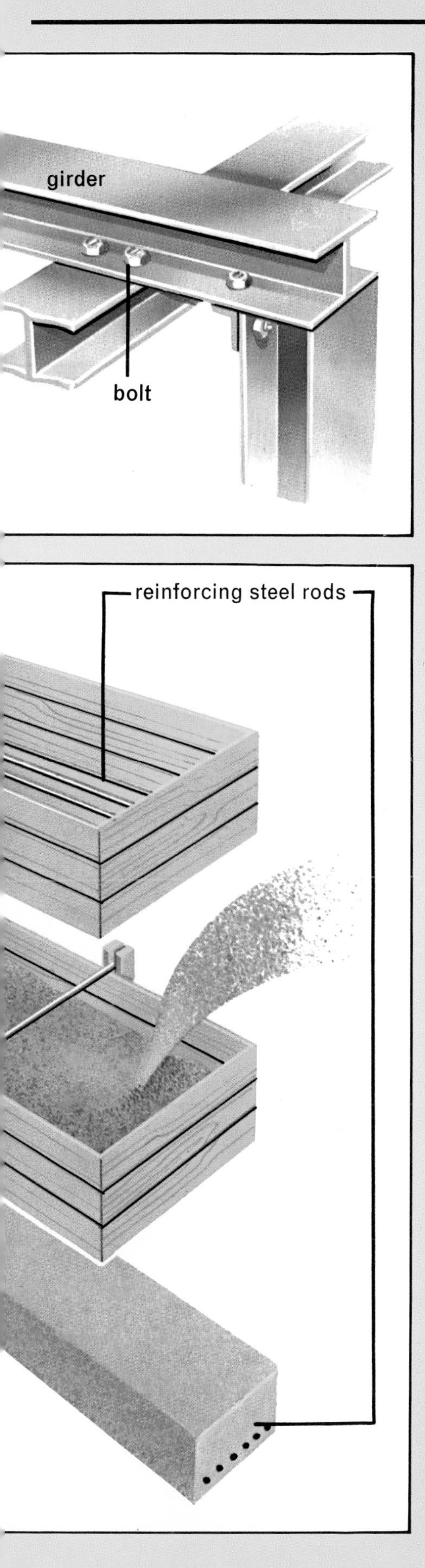

Pre-stressed concrete

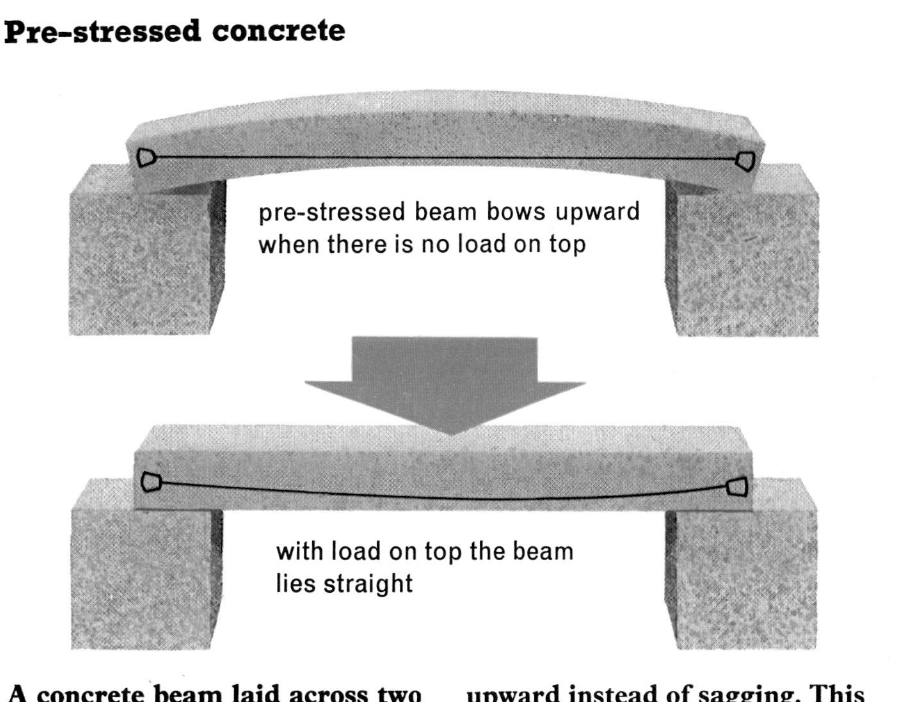

A concrete beam laid across two supports tends to sag in the middle. If an extra weight is added the beam might even break. To prevent this, wires are put into the concrete when it is being cast, and stretched very tight. The beam then bows upward instead of sagging. This technique is called *pre-stressing*. When a weight is added to a pre-stressed beam, the beam sinks to a level shape, and doesn't have to be made bigger and stronger to be able to carry any extra weight.

Lumber

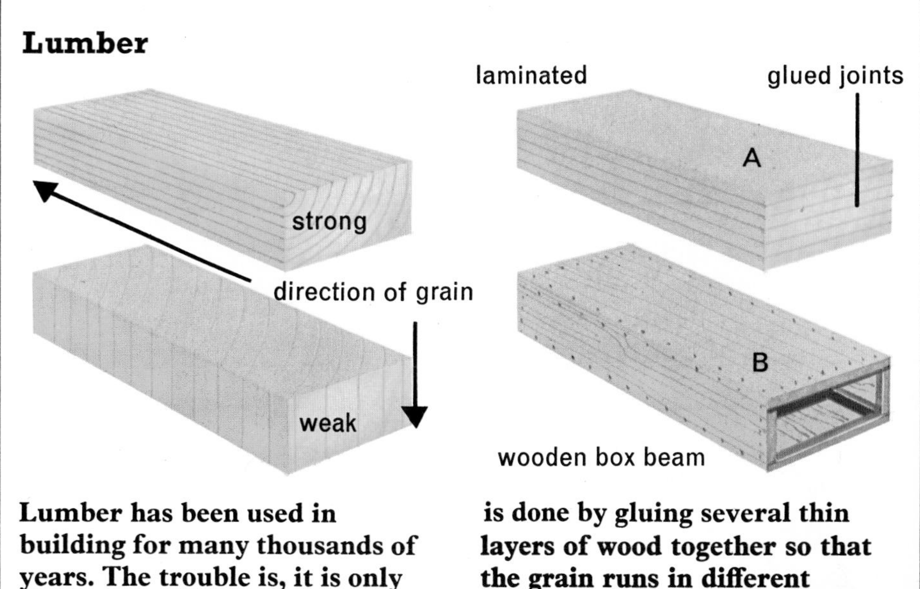

Lumber has been used in building for many thousands of years. The trouble is, it is only strong in one direction—along the grain. Laminating wood helps strengthen it (A). This is done by gluing several thin layers of wood together so that the grain runs in different directions in each layer. Making wood into *box section beams* also strengthens it (B).

Groundwork

Before anything can be built, the ground usually has to be prepared. Uneven ground has to be smoothed out so that the structure can be built on a level surface. Foundations need solid soil to rest on and that is usually found deep underneath the looser topsoil. Vast quantities of this loose earth have to be cleared away by bulldozers, scrapers and excavators.

A machine giant

An excavator is like a best friend to a modern construction engineer, because it has so many uses. Small excavators tackle jobs like digging trenches, while larger machines handle the bigger jobs, such as digging the foundations of a large building. Most excavators have a long jointed 'arm' with a bucket on the end controlled by hydraulic jacks. The bucket has teeth on the edge to scoop out the soil. When the bucket is full, the top part of the excavator turns around and tips the soil into a waiting truck.

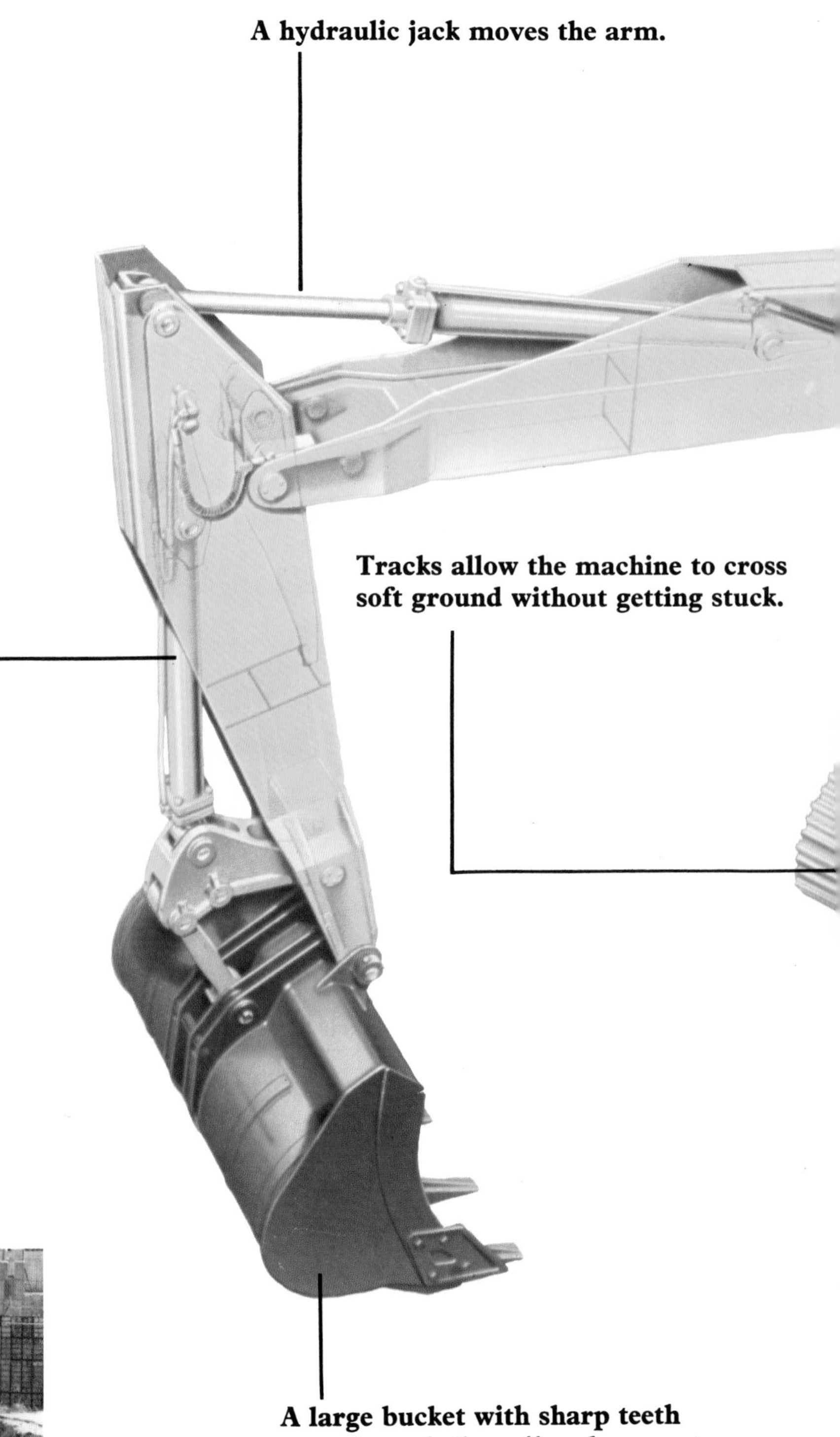

Two tracked excavators pile up the soil before it is removed from the site.

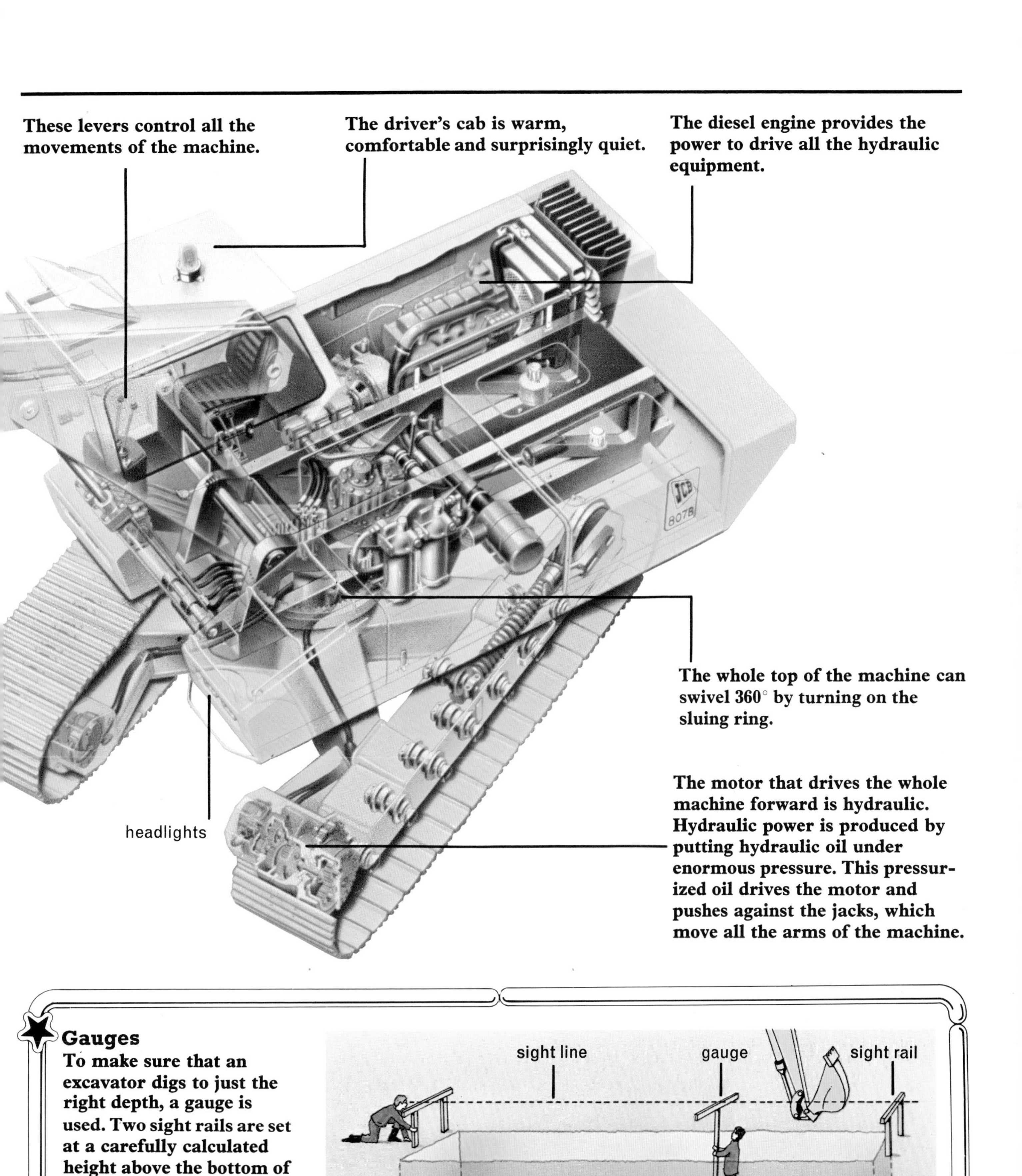

These levers control all the movements of the machine.

The driver's cab is warm, comfortable and surprisingly quiet.

The diesel engine provides the power to drive all the hydraulic equipment.

The whole top of the machine can swivel 360° by turning on the sluing ring.

The motor that drives the whole machine forward is hydraulic. Hydraulic power is produced by putting hydraulic oil under enormous pressure. This pressurized oil drives the motor and pushes against the jacks, which move all the arms of the machine.

Gauges

To make sure that an excavator digs to just the right depth, a gauge is used. Two sight rails are set at a carefully calculated height above the bottom of the trench. When the hole is deep enough, the top of the gauge lines up with the rails.

Laying the foundations

If you have ever tried digging a garden, you will know that the ground can be very hard in some places and very soft in others. Yet houses can be built perfectly safely on either type of soil. How can this be?

No matter what type of ground a building rests on, a foundation is always necessary. If there is very strong rock directly below the loose topsoil, the ground simply needs to be cleaned and leveled before construction begins. The rock itself forms the foundation. But such cases are rare. Usually a foundation has to be specially built.

What is a foundation?

Even very tall structures can be built on weak soil if the foundation goes deep down into the ground or spreads out a long way. It works like this. Imagine you are trying to walk in deep, soft snow. It is very difficult, because with each stride, you sink right down, perhaps up to your knees. But wearing snowshoes, which look something like tennis rackets, there is no problem. Because the snowshoes are bigger and wider than your feet, you won't sink. It is exactly the same with a foundation. The weight of the building is spread out over a wider area and this stops it from sinking.

A different type of foundation has rods or *piles* driven deep down into the ground. It does not rely on spreading the load. If you like, think of a pile as a stick being pushed into mud. At first it goes in easily. Then it becomes harder to push and finally it stops. The stick will support your weight if you stand on it. That is exactly the same job that piles do when they are put underneath a building.

Box foundation
A box foundation takes up less space than a raft foundation. It works somewhat like a raft, because it 'floats' in the soil. A concrete hole is formed in the ground and the structure is built on top of it. The space inside a box foundation can be used as a basement or garage.

Raft foundation
If this tall building sat directly on the soil without a foundation, it would sink. That's because the structure is tall and heavy with only a very small base. To spread the load, a flat slab of concrete has been built below the tower. The weight of the building cannot push the slab through the ground.

Pile foundation
Each steel tube is a solid pile of concrete and reinforcing steel. The workmen will now

build a concrete slab resting on these piles to form the base of the building.

If you could see through the soil, this is what a piled foundation would look like. The piles support the base of the skyscraper.

National Westminster Tower

Right in the heart of the City of London stands the tallest building in Britain—the National Westminster Tower. It rises from a thin column of concrete rather like a tree trunk, and only spreads out 40 m (130 ft) above ground.

A concrete raft

Office blocks are packed tight in the 'City', so the enormous tower had to be built on a tiny site. Beneath the site was clay that would not support the weight of the structure. So 375 piles were driven down into the soil to reach the gravel far below. A huge circular raft of concrete, as big as a soccer field, was cast to join the tops of the piles. Then slowly but surely the column of concrete started to rise.

Higher and higher

High above the ground three great 'leaves' of reinforced concrete were cast onto the main core of the building to support offices built above them. A special concrete-molding machine called a *climbing form* built the rest of the tower. It moved slowly but steadily up the wall, forming the concrete as it went.

Four different groups of men worked at various heights of the tower as it grew skyward. Always at the top were the concreters making the core. Below them were workmen fixing the steel girders for the offices carried on the arms of the building. Then came the men putting in the floors, and just behind them the window fixers. Although the tower took a total of seven years to build, it grew from 45 m (148 ft) to its full height of 183 m (600 ft) in less than one year.

Daily, the tower seemed to grow higher, dominating the skyline.

A dramatic shot looking up inside the core at the climbing form. The three steel columns you can see are cranes.

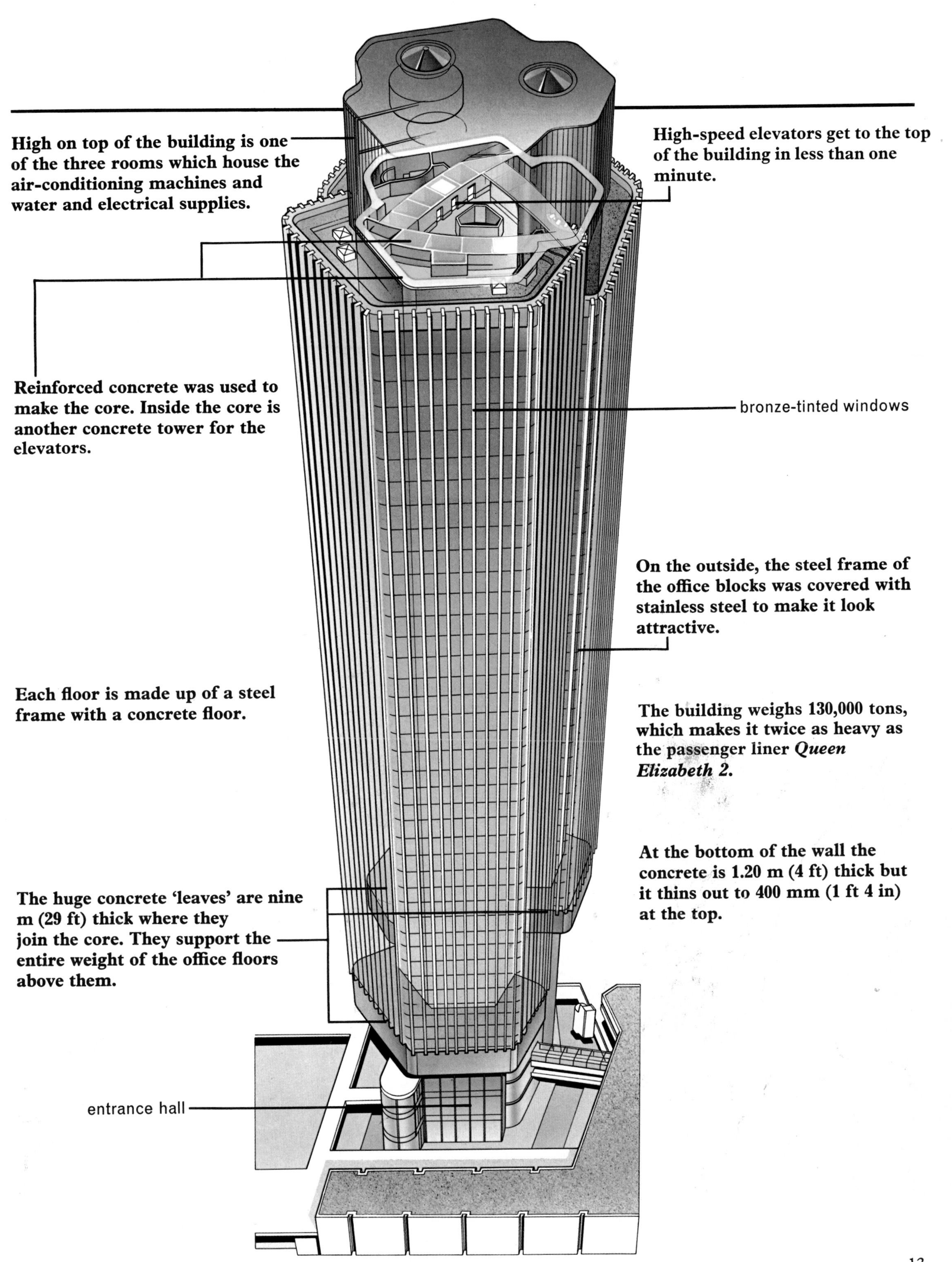
High on top of the building is one of the three rooms which house the air-conditioning machines and water and electrical supplies.
High-speed elevators get to the top of the building in less than one minute.
Reinforced concrete was used to make the core. Inside the core is another concrete tower for the elevators.
bronze-tinted windows
On the outside, the steel frame of the office blocks was covered with stainless steel to make it look attractive.
Each floor is made up of a steel frame with a concrete floor.
The building weighs 130,000 tons, which makes it twice as heavy as the passenger liner *Queen Elizabeth 2.*
At the bottom of the wall the concrete is 1.20 m (4 ft) thick but it thins out to 400 mm (1 ft 4 in) at the top.
The huge concrete 'leaves' are nine m (29 ft) thick where they join the core. They support the entire weight of the office floors above them.
entrance hall

Sydney Opera House

Sydney Opera House in Australia is one of the most complicated structures ever built. The architects had to revise their designs many times before the engineers could start building.

The first stages

The Opera House sits on a small block of rock jutting out into Sydney Harbour. Because the site is so small, many of the concrete parts for the building had to be made in a *pre-casting yard* nearby. From the yard, the completed parts were transported to the site and lifted into place by crane. The massive base of the building was built containing the maze of rooms that would one day be offices and the performers' dressing rooms. Even the seats were cast into the base.

Erecting the roof ★ was a very difficult job for the engineers. By the time they had worked out how to attach all the parts, the shape of the roof was rather different from the first idea of the architect. In all, 2400 pieces were needed to form the ten sections of the roof.

The finishing touch

Once the roof framework was complete, the building looked like a skeleton of concrete arches and ribs. To finish it off, hundreds of concrete panels covered with white tiles were put on top.

In 1973, some 15 years after construction began, Queen Elizabeth II officially declared the building open. It remains a unique and dazzlingly beautiful landmark.

★ Adding the roof

Each arch or 'rib' of the roof was made of several precast sections which were lifted into place by crane. To support each arch before it was finished, a steel frame was built. When the two parts of the arch were securely in place, the steel frame could be moved to the next position. Each arch was joined at its highest point to the one next to it by a concrete beam.

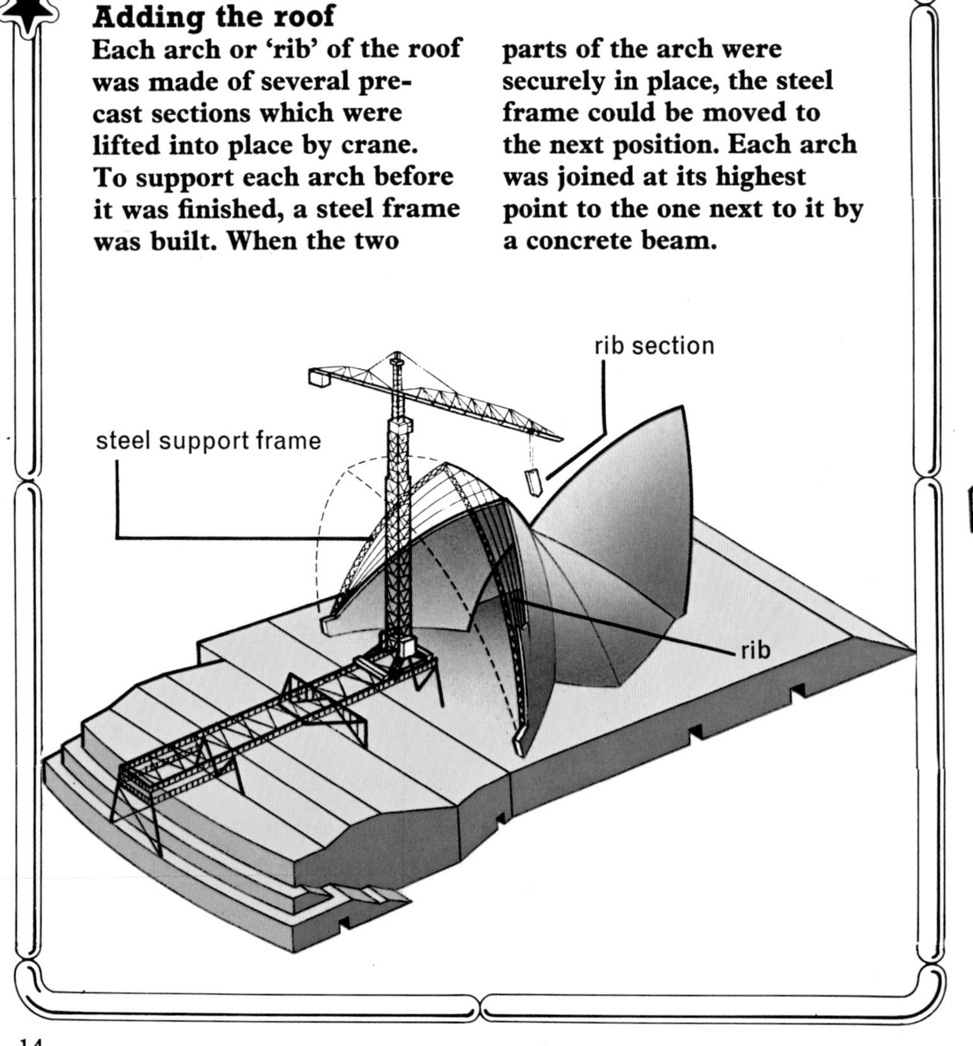

Underneath each shell is a maze of concrete walls which form rooms and theaters.

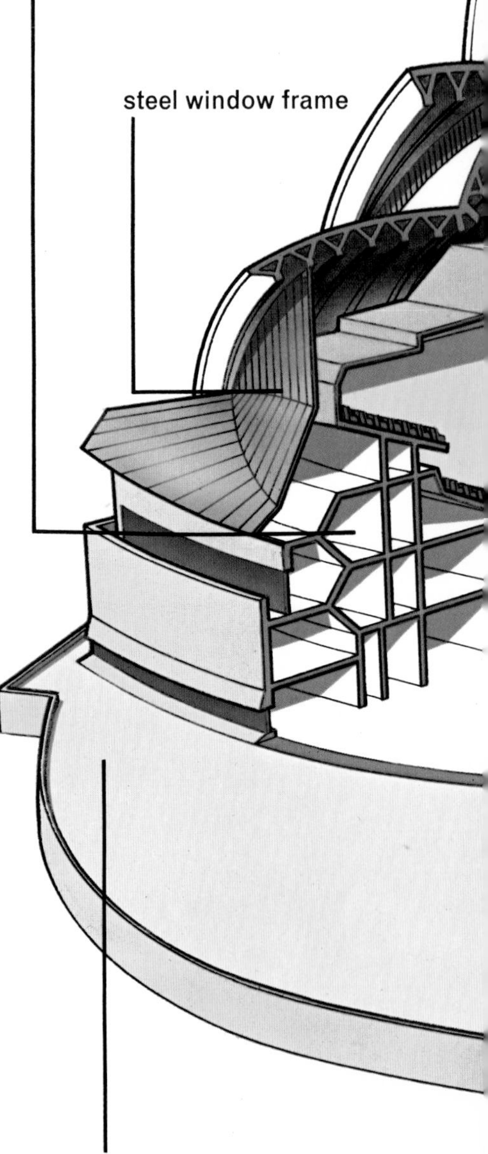

Massive concrete foundations cover the natural rock.

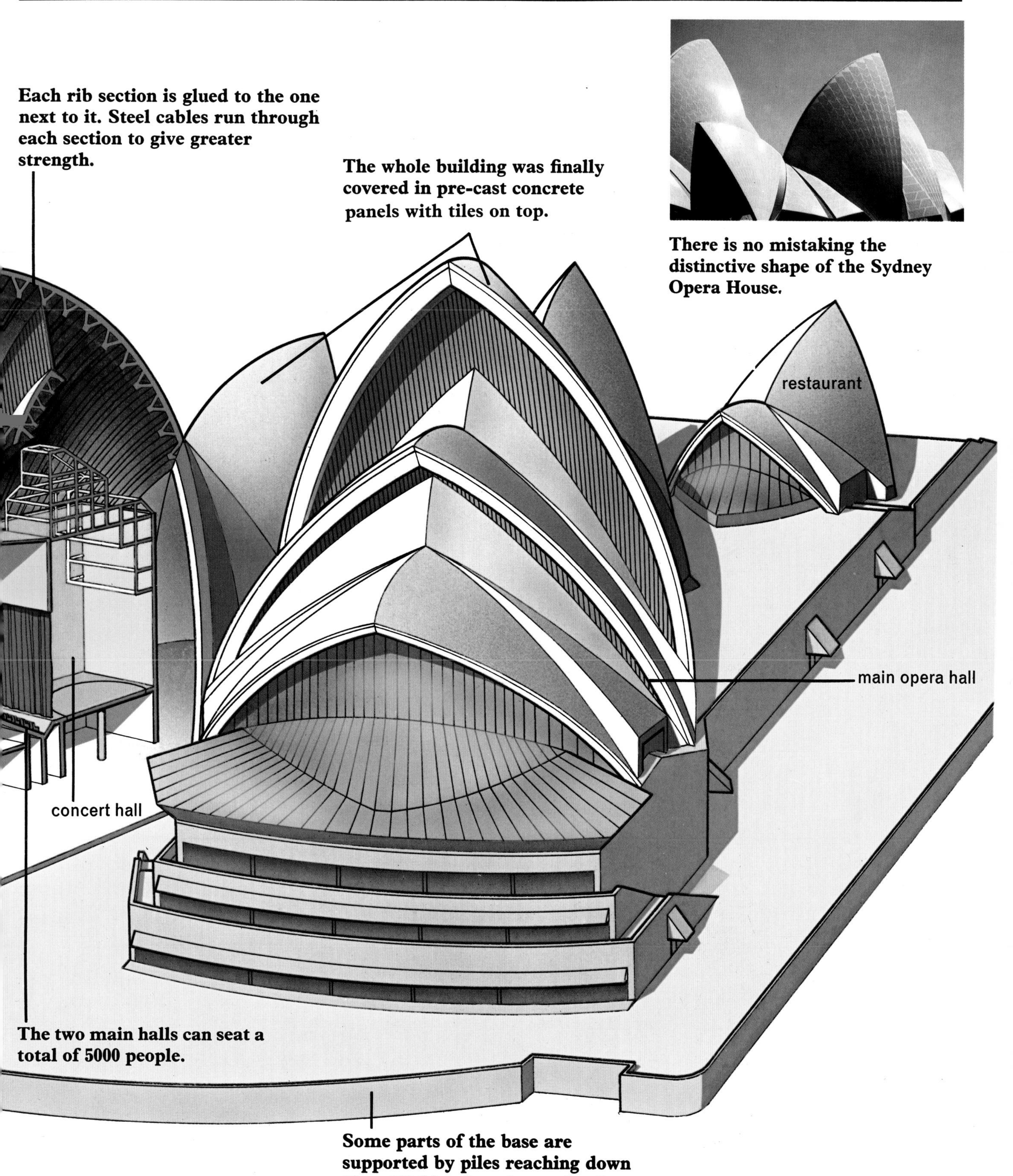

There is no mistaking the distinctive shape of the Sydney Opera House.

Each rib section is glued to the one next to it. Steel cables run through each section to give greater strength.

The whole building was finally covered in pre-cast concrete panels with tiles on top.

The two main halls can seat a total of 5000 people.

Some parts of the base are supported by piles reaching down 12 m (40 ft) below the sea level.

Empire State Building

Halfway around the world from the Sydney Opera House is the famous Empire State Building in New York. The two buildings are completely different but equally spectacular. The Opera House is complicated and beautiful. The Empire State is simple and classic. One took 17 years to design and build, the other 17 months.

Why build skyscrapers?

New York is the home of the skyscraper for two reasons. As in any other big city there is a shortage of room in the city center, so office blocks are built taller and taller to make the most use of the space. The city is also built on very hard rock which allows huge buildings to be erected on simple foundations.

In 1929 the owners of the old buildings on the site of the present skyscraper decided to build an office block which would be the tallest man-made object in the world. A very simple design was worked out to speed up the job, then demolition of the old offices started.

The builders managed to add six new floors every week. Few buildings have ever grown so quickly. Even today, it usually takes much longer. Every day 3500 men went to work on the building and thousands of tons of steel, bricks and stone were delivered. To save time, railroad tracks were laid on every floor to carry materials from delivery elevators to the workmen. At one stage it was taking less than four days to make the steel in the factories at Pittsburgh, form it into girders, take it to New York and fix it in place at the top of the building.

Simple but effective

When it was all finished, in 1931, the project had cost $2 million less than expected and the Empire State Building was hailed by many as one of the wonders of the world.

Surrounded on all sides by skyscrapers, the Empire State Building is still one of New York's most impressive sights.

A disaster

One misty afternoon in 1945 a bomber lost its way over New York and crashed into the side of the Empire State Building at over 400 km/h (250 mph). It was a tragedy no designer could have foreseen. Incredibly, the building withstood the force of the crash.

Tall buildings are designed to withstand many forces. The bottom of a skyscraper has to carry the weight of all the floors above as well as their contents. High winds try to bend the building and snow can add a lot of extra weight. Even the daily rise and fall of temperature makes buildings expand and contract.

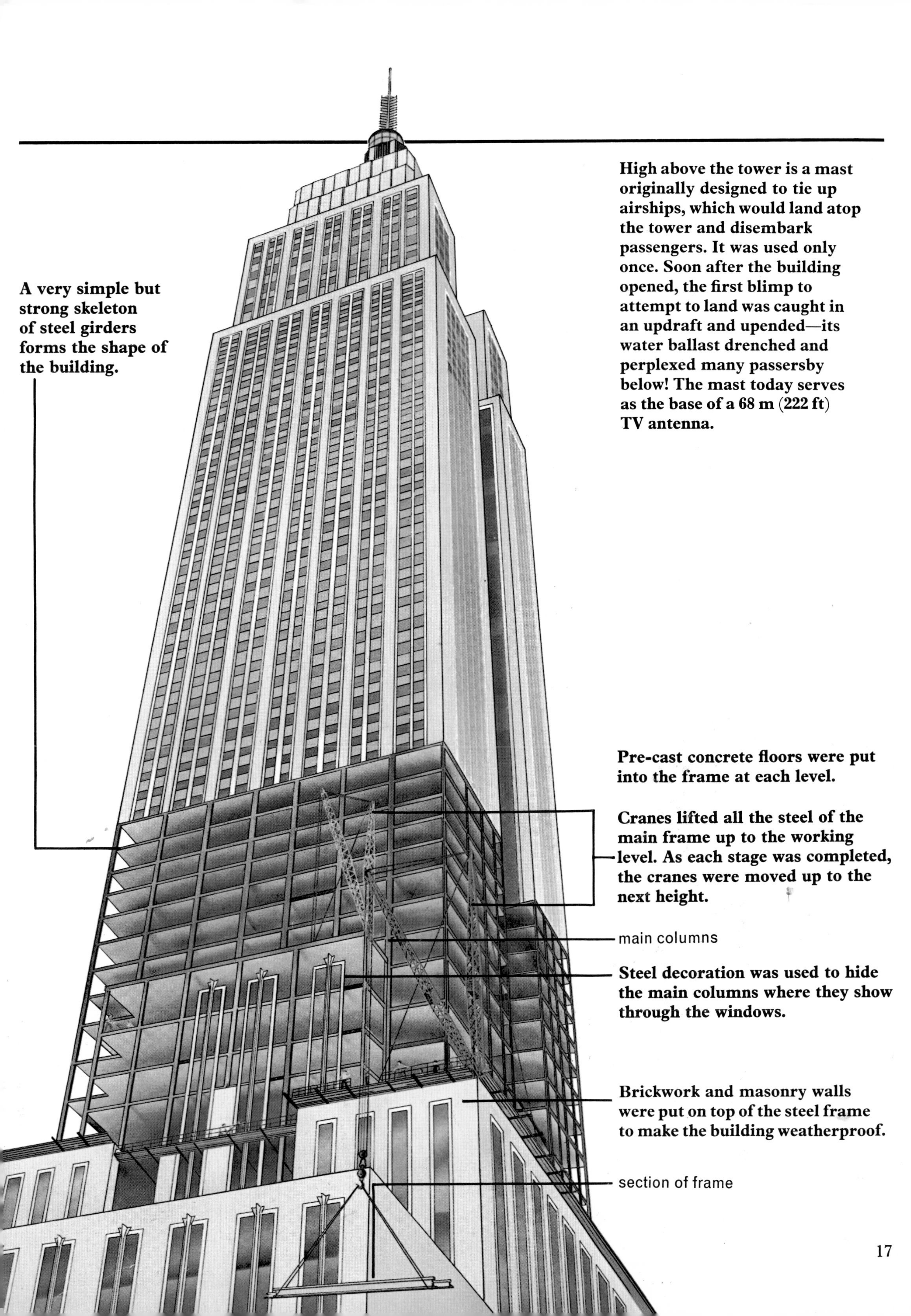
High above the tower is a mast originally designed to tie up airships, which would land atop the tower and disembark passengers. It was used only once. Soon after the building opened, the first blimp to attempt to land was caught in an updraft and upended—its water ballast drenched and perplexed many passersby below! The mast today serves as the base of a 68 m (222 ft) TV antenna.
A very simple but strong skeleton of steel girders forms the shape of the building.
Pre-cast concrete floors were put into the frame at each level.
Cranes lifted all the steel of the main frame up to the working level. As each stage was completed, the cranes were moved up to the next height.
main columns
Steel decoration was used to hide the main columns where they show through the windows.
Brickwork and masonry walls were put on top of the steel frame to make the building weatherproof.
section of frame

Pompidou Center

'When will it be finished?'

Many first-time visitors to the Pompidou Arts Center in Paris ask this question. For although the Center was finished and opened in 1977, it still looks incomplete. From the outside it looks as if the scaffolding has not been removed. But look closer and you will see that it is not scaffolding but the frame of the building, including the stairs and elevators. Inside the building there are no supporting columns in sight, because the load-bearing structures are all on the outside★. The Arts Center is, to say the least, an unusual building.

A solid foundation

The first job for the engineers was to make a suitable foundation. It had to be a box foundation to give room for more floors below ground. Workmen dug a hole 15 m (49 ft) deep, then special supports were lowered into the ground. Soon all the floors up to ground level were added.

Steel from Germany

The vast quantity of steel needed for the frame was assembled and welded at steelworks in Germany. Special rail trucks were used to take the steel to Paris. At the site, enormous cranes

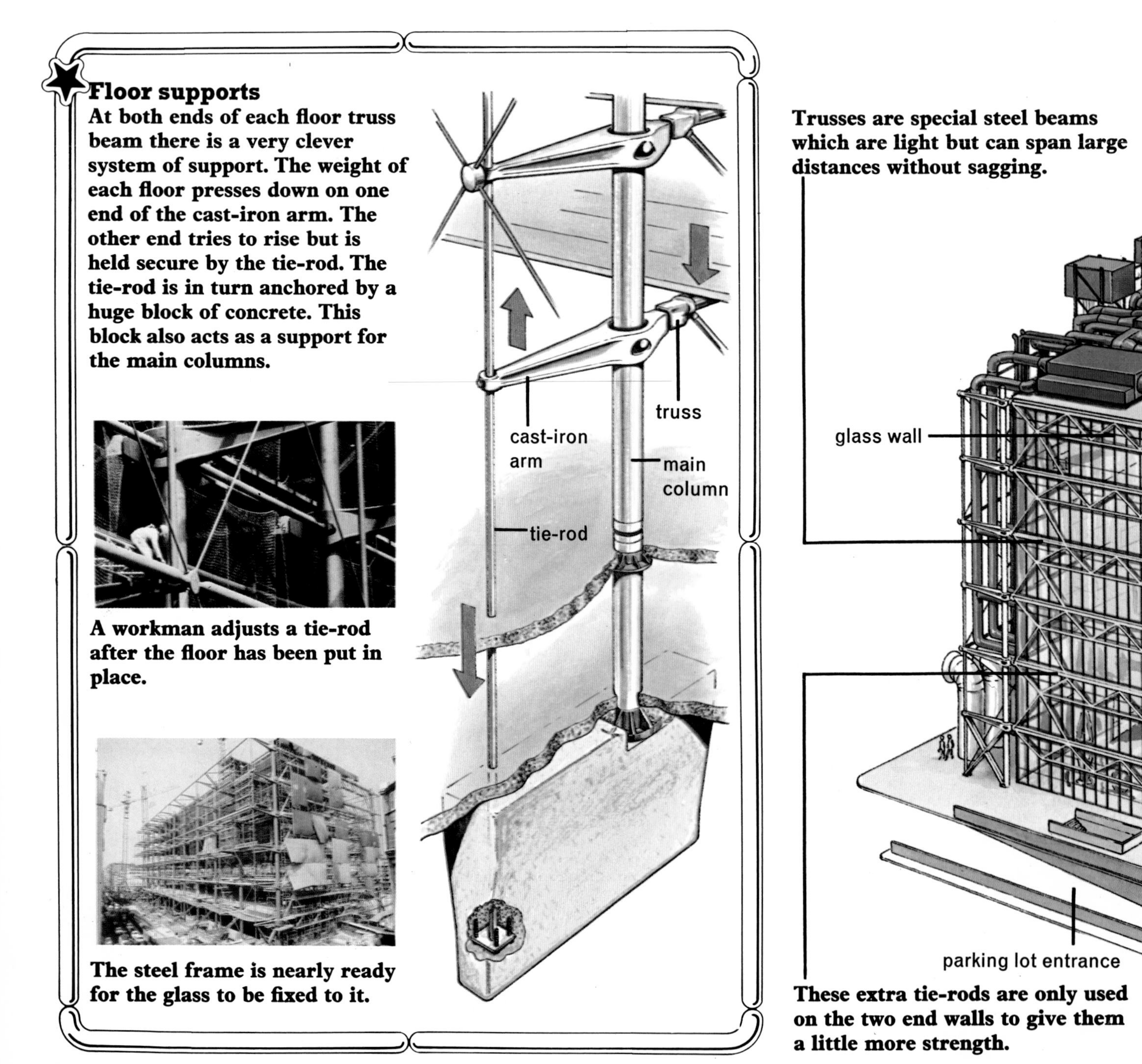

★**Floor supports**

At both ends of each floor truss beam there is a very clever system of support. The weight of each floor presses down on one end of the cast-iron arm. The other end tries to rise but is held secure by the tie-rod. The tie-rod is in turn anchored by a huge block of concrete. This block also acts as a support for the main columns.

A workman adjusts a tie-rod after the floor has been put in place.

The steel frame is nearly ready for the glass to be fixed to it.

Trusses are special steel beams which are light but can span large distances without sagging.

These extra tie-rods are only used on the two end walls to give them a little more strength.

hoisted the heavy framework into place. Then all the remaining floors could be added.

The building was not yet complete. To supply electricity, water, air-conditioning and telephones, a complicated system of tubes and ducts was attached to the frame. The glass walls that would normally form the outside of the building are just visible beneath this dense network.

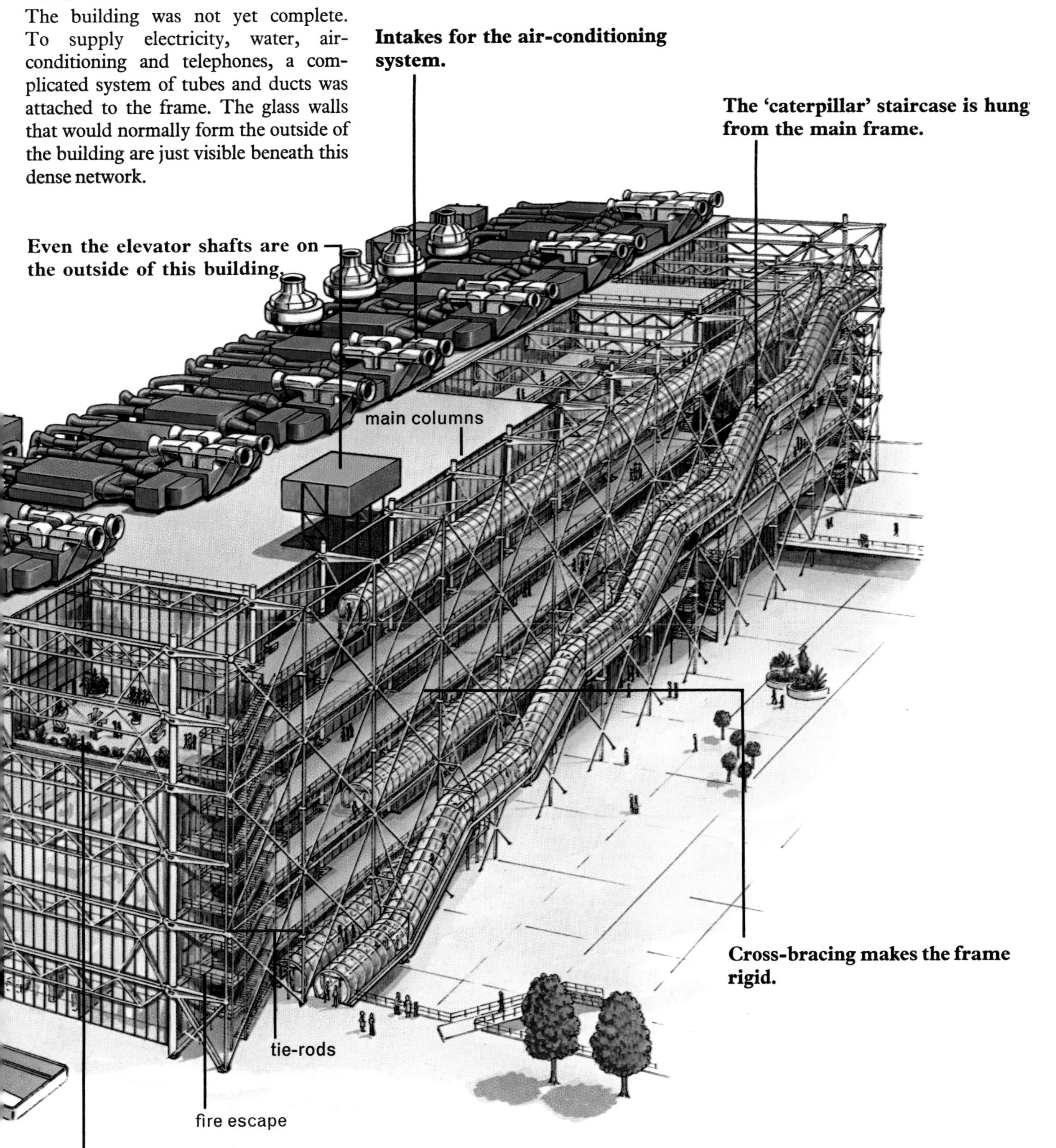

Intakes for the air-conditioning system.

The 'caterpillar' staircase is hung from the main frame.

Even the elevator shafts are on the outside of this building.

Cross-bracing makes the frame rigid.

Parts of the glass wall have been cut out to form balconies.

Munich Olympic Stadium

Every four years the Olympic Games, the greatest event in the world's sporting calendar, take place. There are so many competitors that whole new Olympic 'villages', as well as new arenas and grandstands, have to be built★. In 1972, when the Games were held in Munich, a very odd-looking athletic stadium was designed, which was a brilliant technical achievement.

There is nothing worse than being stuck behind a pillar when you're trying to watch a race. At Munich there are no pillars to spoil the view. How then is the roof held up?

Tent principle

The roof of the main stadium is a sheet of special plastic attached to leaning steel pillars. The pillars in turn are supported by steel cables anchored to the ground by huge concrete blocks. Cranes had to be used to lift the pillars into place, then the cables, or *stays*, were attached. Slowly the pillars were eased into their leaning positions and more steel cables were hoisted up over the top of the pillars and anchored securely. All the stays were pulled tight and the shape of the roof was formed. It is probably easier to imagine the structure as a giant tent with the steel cables acting as the guy lines. The design of the stadium was a huge success. The roof let in plenty of light yet kept off the rain and, most important of all, with all the supports above or behind the spectators, everyone had an excellent view.

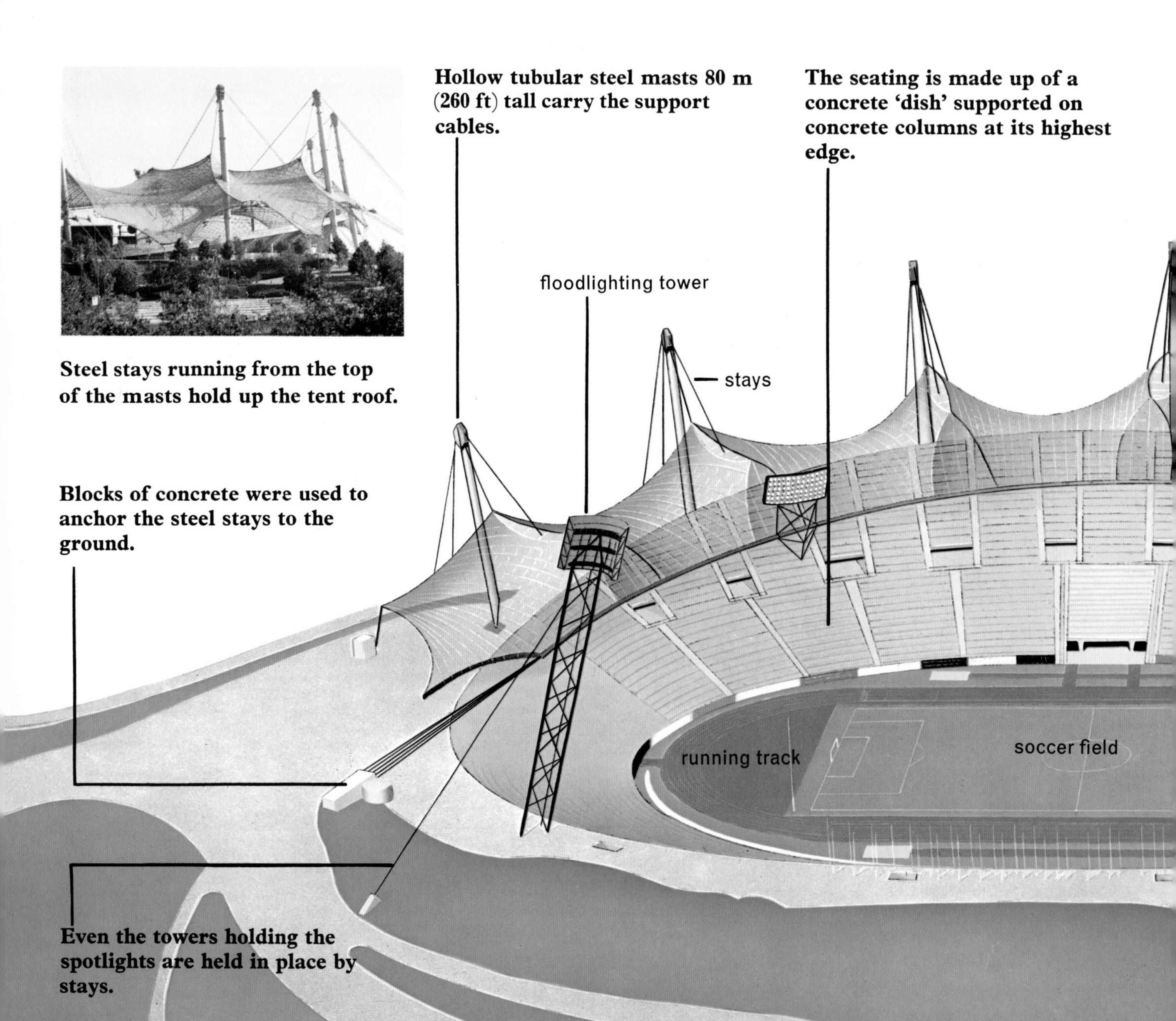

Steel stays running from the top of the masts hold up the tent roof.

Hollow tubular steel masts 80 m (260 ft) tall carry the support cables.

The seating is made up of a concrete 'dish' supported on concrete columns at its highest edge.

Blocks of concrete were used to anchor the steel stays to the ground.

Even the towers holding the spotlights are held in place by stays.

Years of planning

The Olympic Games are held every four years in a different country. It is a great honor to be the host country and a lot of effort is put into planning a spectacular stadium. The picture on the right shows the stadium being built in Moscow for the 1980 Olympics. It costs a lot of money to build sports stadia.

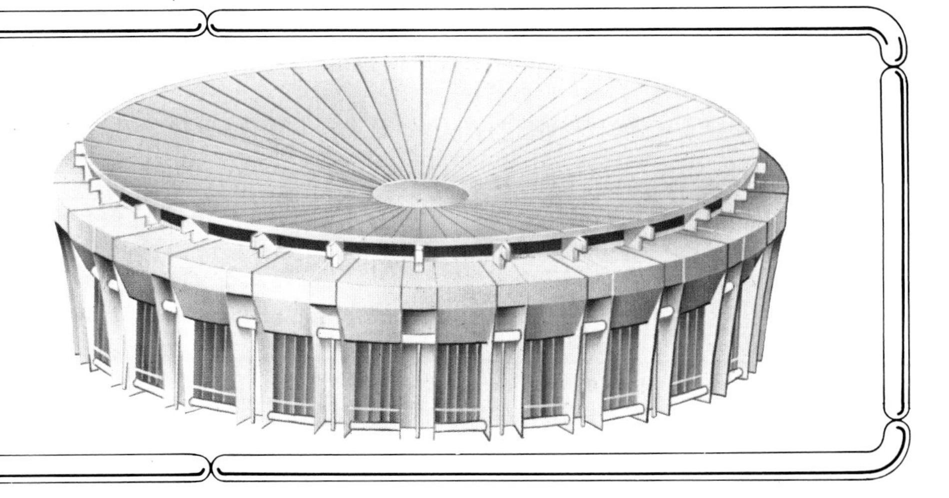

A multi-strand cable, attached to two huge blocks of concrete, runs around the front edge of the roof. The masts pull the roof back to rest against this cable.

Each mast sits on a cleverly constructed ball-shaped joint. This joint allows the mast to lean at just the right angle.

The thousands of clear plastic sheets which make up the Munich stadium roof are joined together by steel strips.

These stays hold up the mast and the roof at the same time.

clear plastic roof

The ground around the stadium has undergone great changes. Landscape gardeners have planted trees and shrubs and added two lakes and an artificial hill.

UN City

A whole new city is being built just outside Vienna, the capital of Austria. It will be another headquarters of the United Nations, in addition to New York and Geneva.

At first glance the collection of buildings in this space-age city looks quite ordinary. Yet the towers, which look Y-shaped from above, were very complicated to build.

Taking the strain

In the middle of each block is an elevator shaft designed to support only the elevators. All of the floors are suspended from large concrete columns at each 'corner'. At three different levels there is a special floor, called a load-bearing floor★, which holds up as many as 13 floors above it. It is a long, delicate operation to lift these floors up in one piece to the correct position.

When complete, the massive tower blocks are extremely heavy. Their great weight has caused each of the buildings to sink slightly into the ground. The elevator shaft, which weighs much less, does not sink quite so much. Special bearings are therefore built into the elevator shaft so it can be lowered to the same level as the other parts of the tower.

The conference hall

Right in the center of the city is the circular conference hall. From the inside it is difficult to tell how the floors are being held up. The answer is that all the floors hang from the roof by a huge steel column which does not reach down to the ground.

A city to be proud of

The UN complex stretches over a vast area. Yet the Austrian government is so proud that Vienna was chosen as the site for the city that it is going to charge a rent of only one Austrian schilling per year. That's about 7 cents!

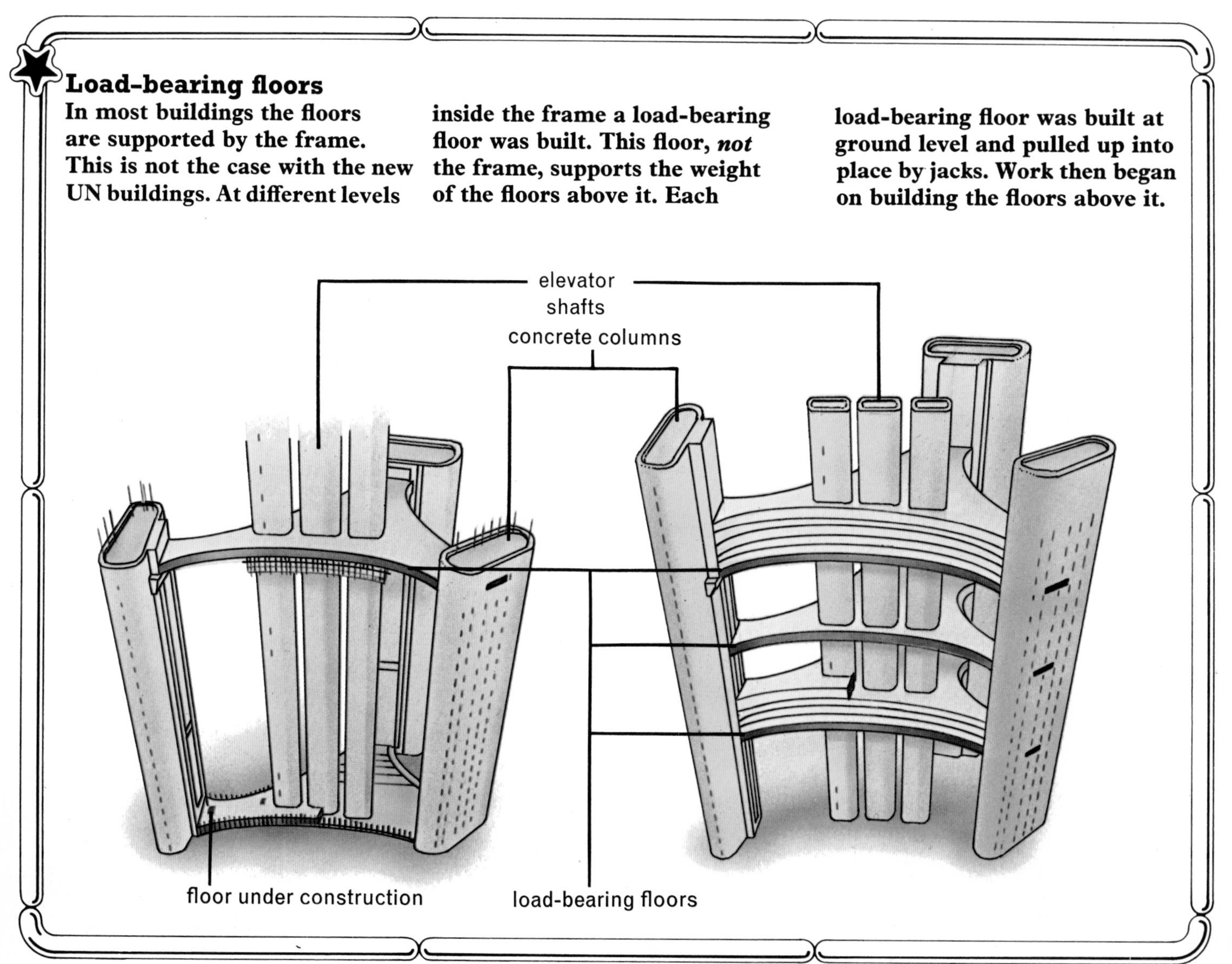

★ Load-bearing floors

In most buildings the floors are supported by the frame. This is not the case with the new UN buildings. At different levels inside the frame a load-bearing floor was built. This floor, *not* the frame, supports the weight of the floors above it. Each load-bearing floor was built at ground level and pulled up into place by jacks. Work then began on building the floors above it.

New floors are being built on several different levels at once. Floors at the top are often ready before floors lower down.

As the conference hall, below, neared completion, work started on the entrance area. The designs drawn up were for an extremely simple concrete frame.

Domes

A dome looks rather like a bowl turned upside down. The next thing to notice is that it needs no supports to hold it up in the middle. In its simplest form a dome is very easy to build.

An ancient art

For thousands of years, people have lived in dome-shaped dwellings. Even today, Eskimos use this design when they build their igloos from blocks of snow. Tribesmen in some parts of Africa plant several long, easily bent sticks into the ground to form a circle, then bend them into the middle and tie them together at the top. Animal skin or some other material is then put over the sticks.

Modern domes

Robert Buckminster-Fuller, an American designer, has been building domes for many years. He designed and built the dome for the U.S. Pavilion at Expo '67 in Montreal. Because of the dome's light weight, its foundations were small. All the individual sections of the dome were made to the right size before arriving on site. A team of men working from scaffolding quickly built up the lattice of steel tubes that form the frame. It was then just a matter of bolting together the tubes and connectors, and a dome 76 m (250 ft) across was complete and ready for use.

Make a paper dome

You will need:
Thin cardboard or stiff paper
Scissors
Pin or compass
Modeling knife
Pencil
Ruler
Quick-setting glue

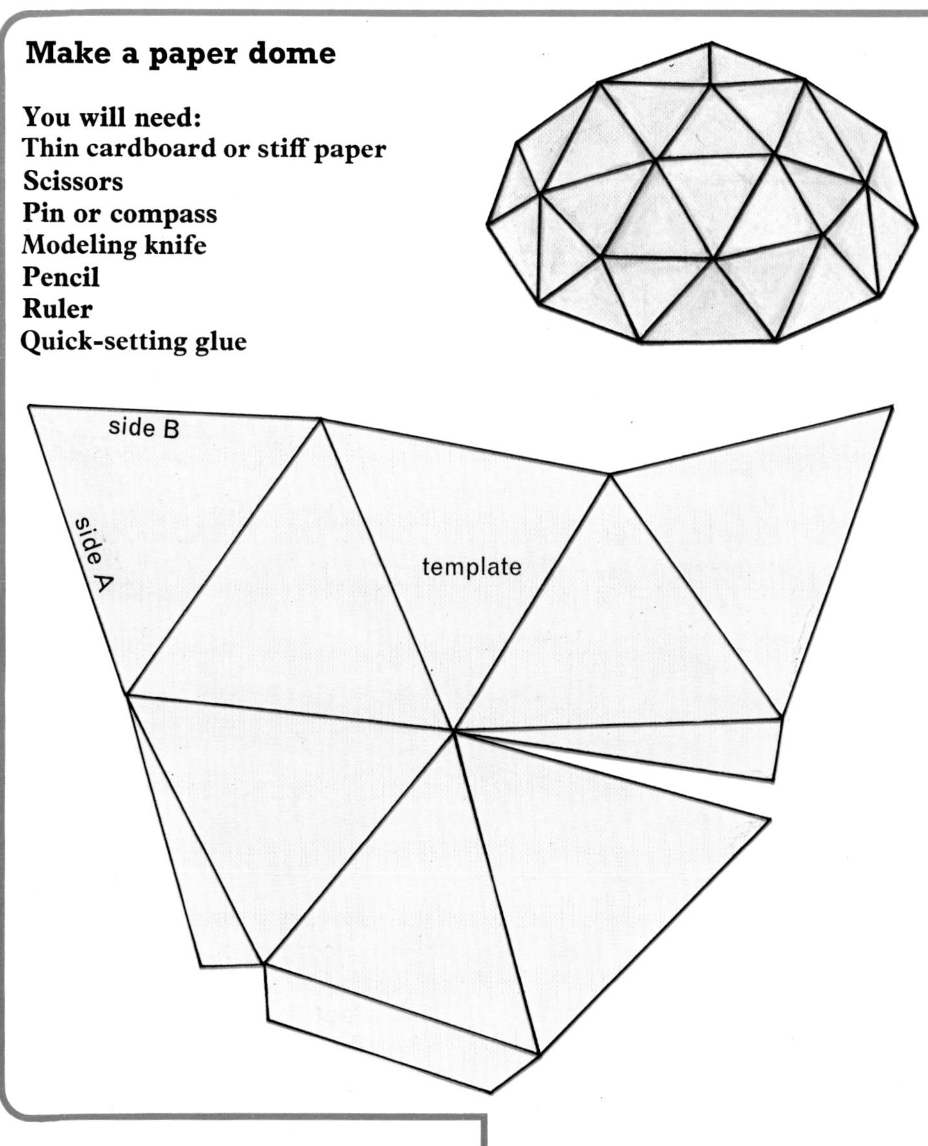

The best-known dome designed by Buckminster-Fuller is this one for the U.S. Pavilion at the Montreal International Exhibition.

This dome is similar in pattern to the one you can make with straws shown on the right.

Make a dome from straws

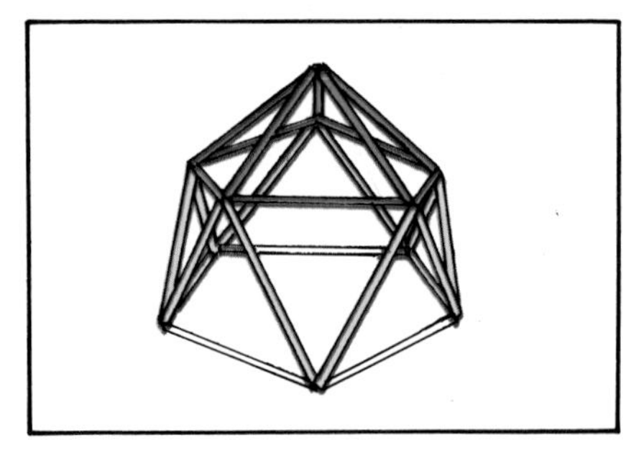

You will need:
Modeling knife
Dressmaker's pins
Pencil erasers
Plastic drinking straws: 5 white, 5 red, 5 blue and 10 yellow

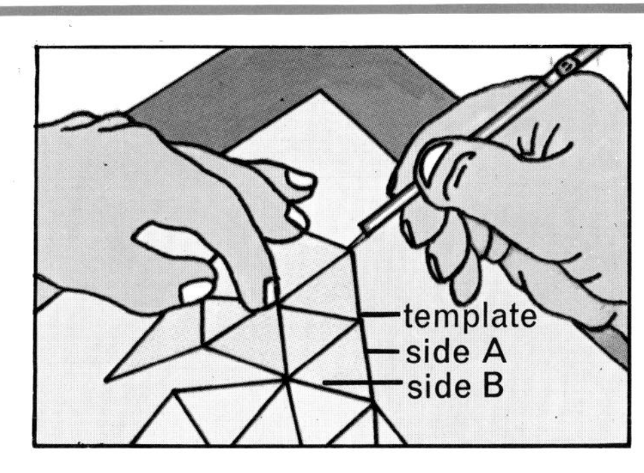

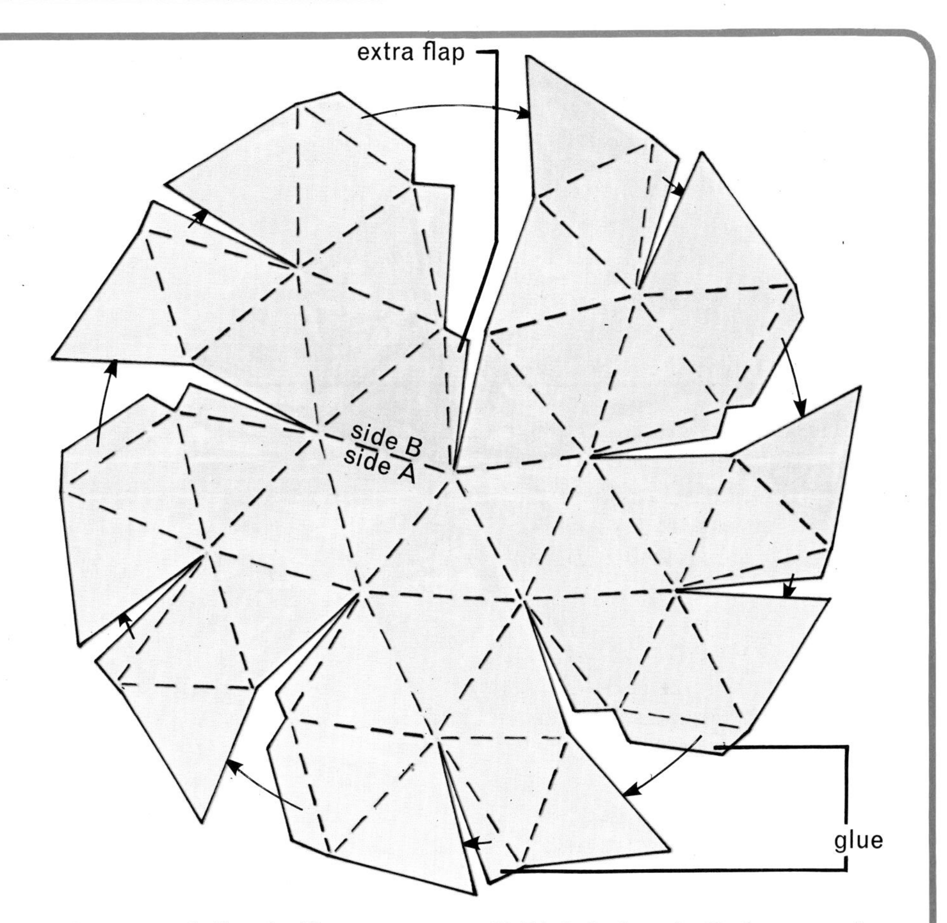

1. Trace the shape shown on the left onto some cardboard. This is the template. Cut it out taking care to include all the flaps. Mark sides A and B.
2. Lay the template on the square of cardboard and, holding it down firmly, use the pin to prick around the shape onto the cardboard underneath.
3. Using the pencil and ruler, join up the pin pricks to form the first section. Mark sides A and B as in the diagrams.
4. Take the template again and lay side A exactly against side B of the section you have already drawn. Again prick through and join up the dots. Repeat the process until you have five sections. Note that the last section does not join onto the first, and has an extra flap.
5. Using a modeling knife, *lightly* score *all* the lines shown as dotted on the diagram.
6. Cut out the finished pattern around the lines shown as solid on the diagram.
7. Lightly bend all the scored lines downward.
8. Glue one set of flaps at a time, firmly pressing the flaps *underneath* the joining edge. Your dome will soon take shape.

1. To make the joints, flatten the ends of the straws to be joined and push a pin through them. Cut pieces of eraser and push a piece onto the end of the pin.
2. Take the five red straws and join them all at one end. This is the vertex.
3. To make the side panels use five blue straws. Join the loose ends of the red straws to the blue straws. Add two yellow straws to each of these points.
4. To make the base use five white straws. Join up the loose ends of the yellow straws as shown in the drawing.

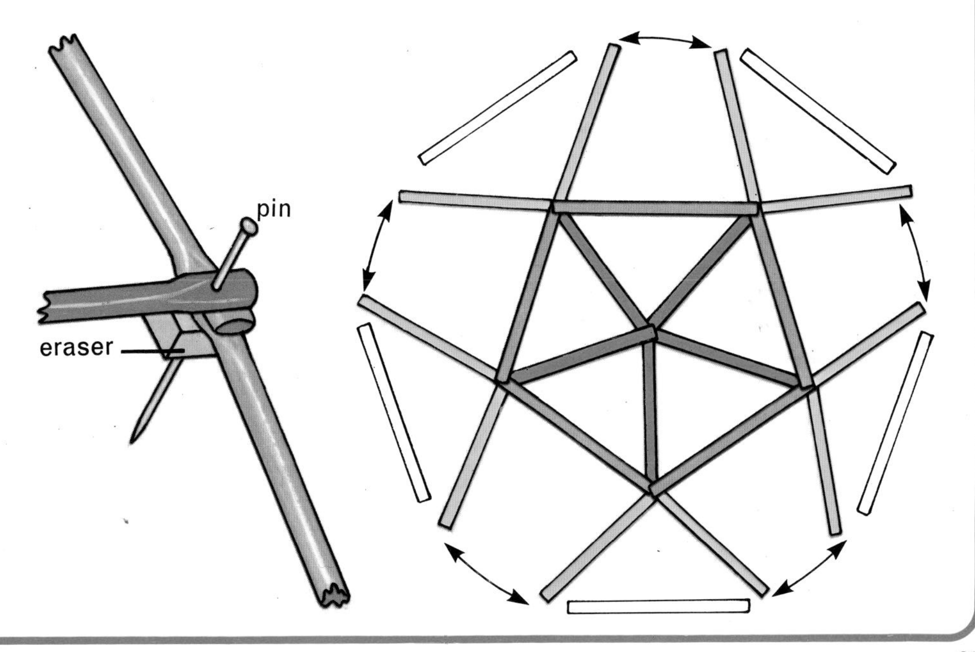

Canadian National Tower

Towering high over the city of Toronto is a slender structure of concrete and steel—the Canadian National Tower. There are taller man-made towers, but all of them are 'stayed' towers, which means that they need cables to hold them steady.

Working round the clock

The first task was to build a firm foundation. Thousands of tons of concrete were poured into a huge hole to form the biggest raft foundation in the world. With that done, the tower grew very fast. Shifts of men worked around the clock to meet schedules. Specially equipped cranes could lift all the building materials up as high as 460 m (1509 ft). Above that height, a huge helicopter had to be used. A massive steel tower was taken up in bits to the top by the helicopter, then bolted together again.

Work and pleasure

Very close to the top a revolving restaurant has been built where people can enjoy a magnificent view while they dine. Finally, high above the restaurant is a powerful radio and television antenna.

For many years the Eiffel Tower was the tallest tower in the world. But today, television masts often reach much higher.

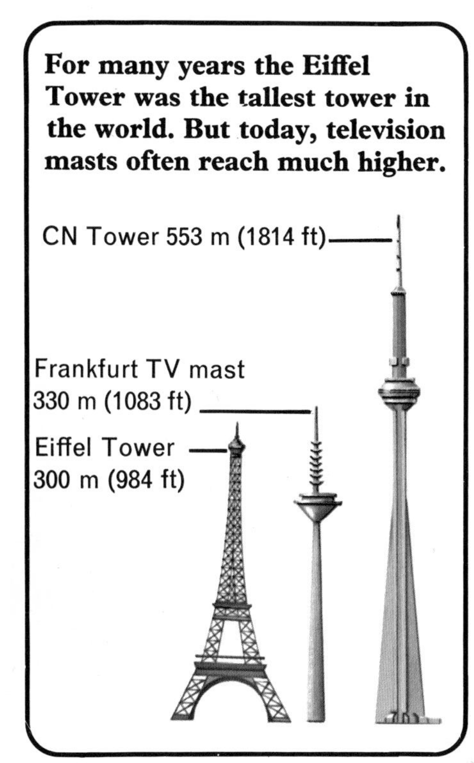

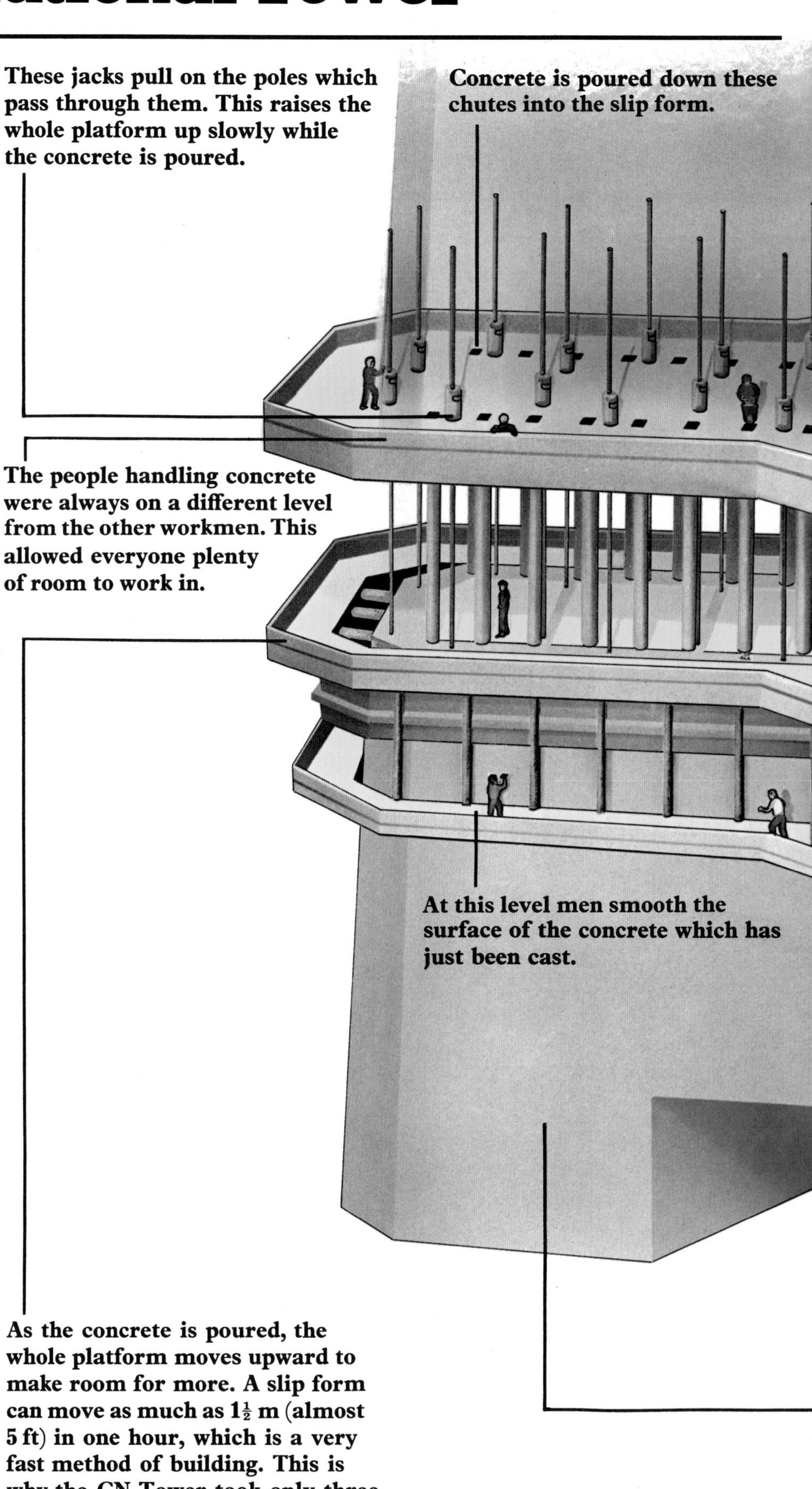

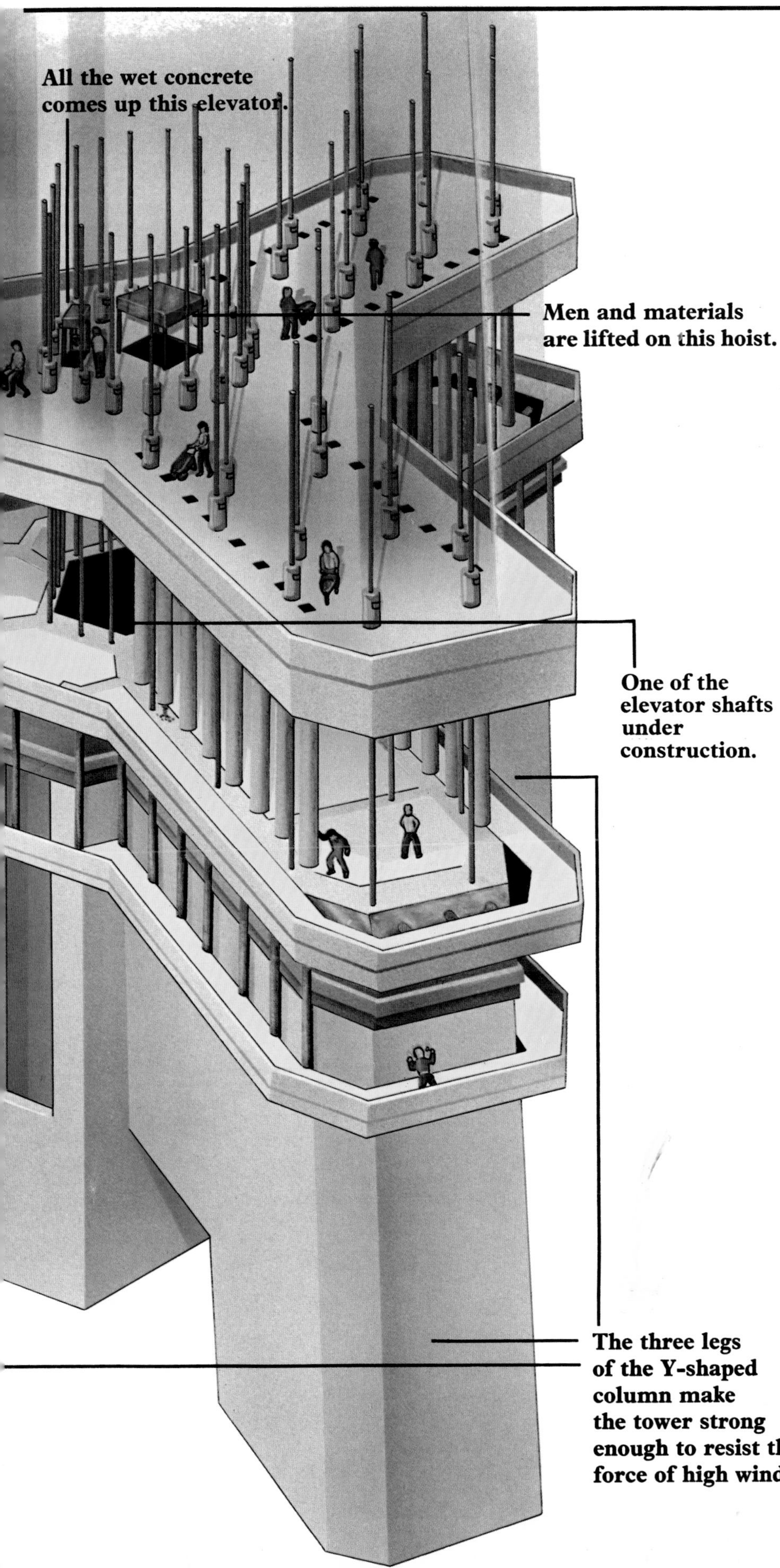

A helicopter lifts the last sections into place. The radio mast right at the top is coated with special material to stop ice from forming.

Even the tallest skyscrapers in Toronto are dwarfed by the giant CN Tower. Just below the top is a revolving restaurant.

Humber Bridge

There is one obvious problem about building a bridge across a wide river or valley. If you build all the supports in the river, it is a difficult job and an expensive one. A clever way around this problem is to hang the bridge from two cables stretched across the river or valley. This is a suspension bridge.

Not according to plan

Construction of the Humber Bridge on England's east coast was an ambitious project. It is the longest suspension bridge in the world. Not everything went smoothly while it was being built. Engineers decided to put one pair of towers on the north shore line and the other pair in the water near the south bank. Using special equipment, the north towers were built from concrete in just 17 weeks. But on the south bank things went wrong. The *caissons* (large watertight, hollow blocks) which were to support the south tower became stuck. It was an enormous task to sink them down into the river bed and then put the tower on top.

Adding the deck

Three kilometers (nearly two miles) down the river, welders made all the parts for the huge steel deck that would eventually hang from the two main cables. At the same time, work began on spinning together the hundreds of steel strands that would make up each main cable★. As each section of the deck was completed, it was steered up-river by tugs to a position below the main cables. All the parts for the deck were then winched up to the correct height and welded together.

★ Cable-spinning

Each of the two mighty cables supporting the roadway is made up of many strands of steel. Each strand is taken across the river by a small wheel hung on a temporary cable. With one end of the strand attached to the anchor block, the wheel is pulled over to the other side, taking the strand with it. At the other side the loop of strand wire is removed and attached to the other anchor block.

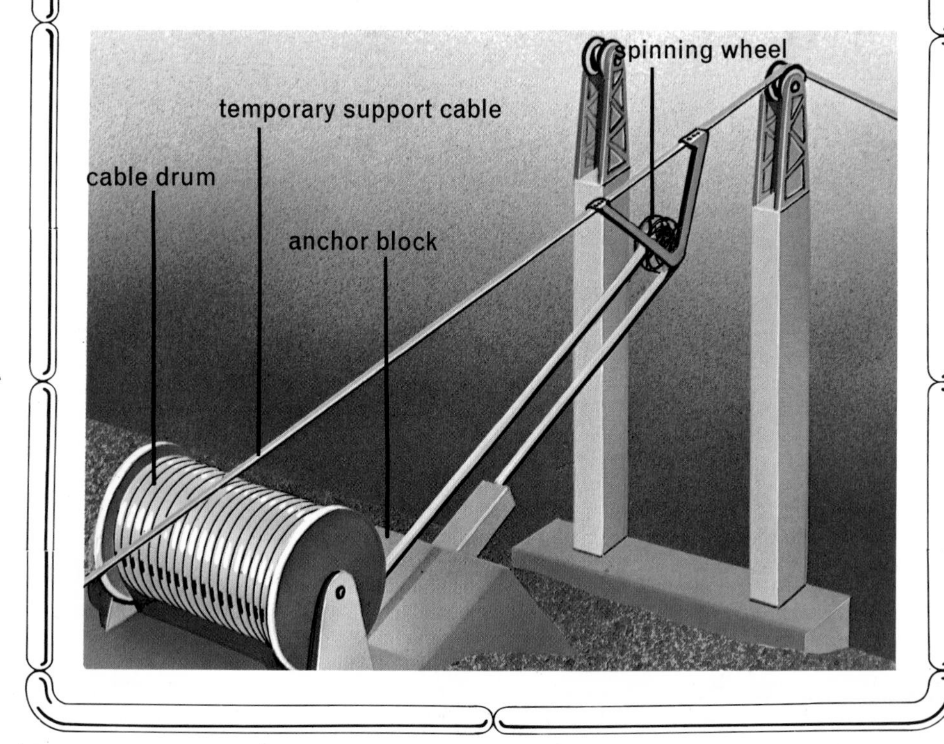

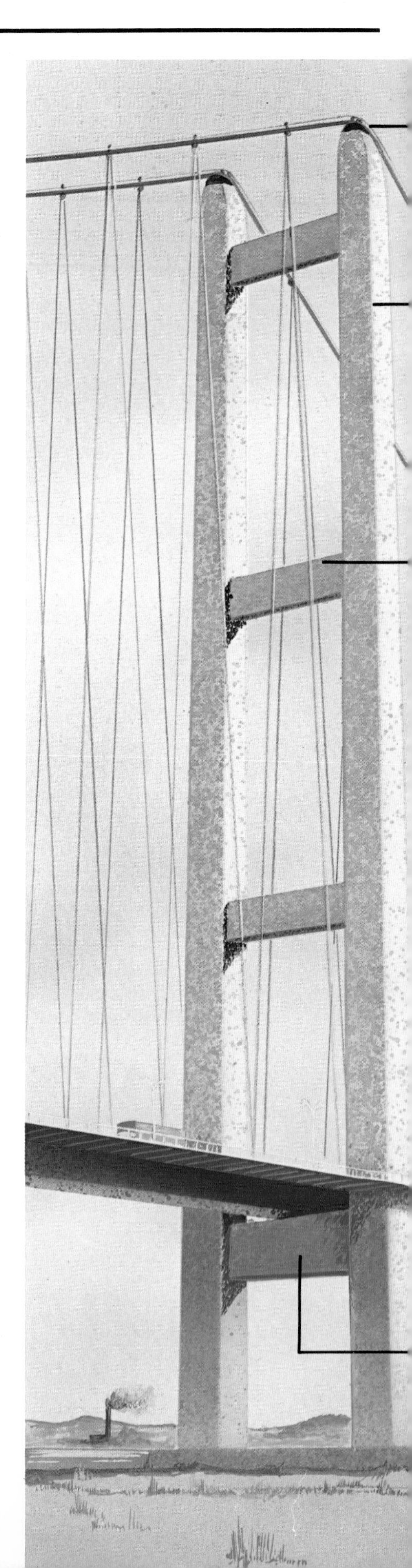

The cables are placed in a steel groove on top of the tower to allow for any slight movement.

Each concrete tower is hollow. Inside, an elevator takes engineers to the top to inspect the cables.

A section of the road deck ready to be added. At each side of the roadway are cycle and footpaths.

Concrete cross-beams stop the towers from leaning away from each other.

Cable-spinning in progress. In the foreground are the concrete blocks where each cable is anchored.

Wires called *suspenders* take the weight of the road deck.

The distance between the two sets of towers is 1380 m (4527 ft).

The deck rests on this beam.

New River Gorge Bridge

It took 150 million years for the New River in West Virginia to carve out a deep gorge. It took just three years to build an arch bridge★ spanning the gap created by the mighty river.

Delayed by bad weather

It was bad luck that construction work began on the bridge during one of the worst winters ever experienced on the east coast. High winds often brought work to a stop. The special crane system used to assemble the steelwork could not carry the huge pieces of the bridge in safety.

Two lots of concrete foundations were built into the sides of the gorge. At the same time, two towers were built on opposite shores. The sections of the bridge, sometimes weighing as much as 86 tons, were carried out and positioned by cranes mounted on cables stretched between these towers. The critical moment came when the last middle section was added. But the calculations were perfect.

Next, the columns could be built and the road deck laid on top. Before long, the highway was open to traffic. Drivers can cross the 300 m (1000 ft) between the banks of the gorge in seconds, saving 45 minutes' drive on the old road.

Sections of the arch were built on the side of the gorge and then lifted into place by a specially built crane. A cable, called a back stay, held each section in place.

When all the sections were in place, the stays could be removed and the arch supported itself. The lattice columns could then be built up to the level of the deck.

Starting from each side of the gorge, the deck beams were laid on the tops of the columns. Concrete slabs filled the spaces between the beams. Then the road was laid.

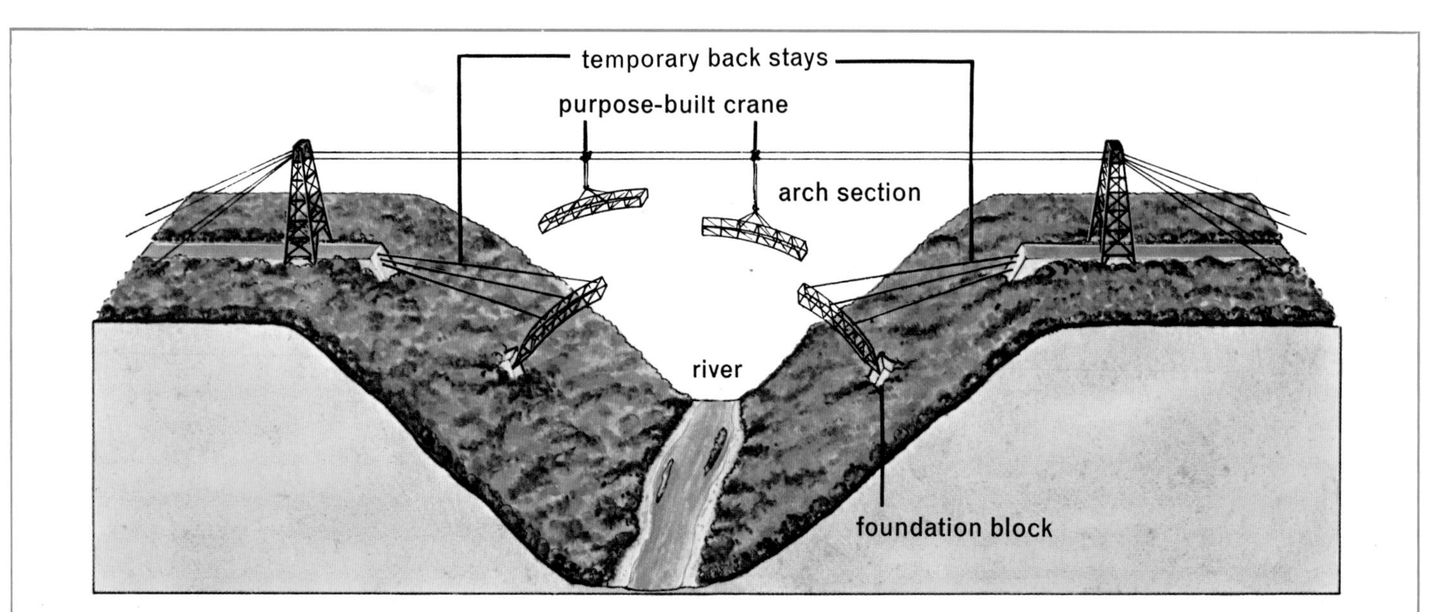

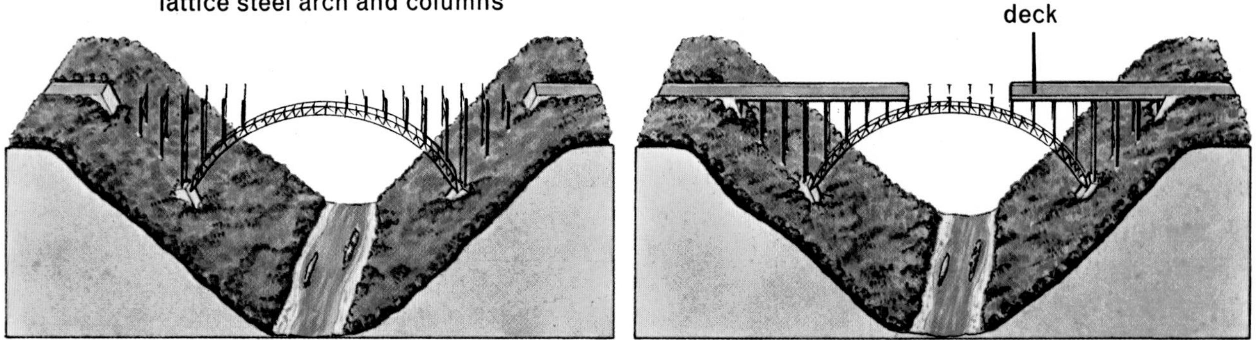

The simple but elegant lines of the steel bridge do not look out of place in the majestic scenery of the New River Gorge.

Lots of thin steel beams were joined together in a criss-cross pattern to make a strong structure which is also very light. This is called lattice-work, and it is frequently used in bridges.

Early bridges

The oldest bridges were flat stones or tree trunks laid across a river. Bridges only became much bigger when people discovered how to build arch shapes. Arches work because the carefully shaped stones at the top are supported by the next ones down and so on right to the ground. As long as there is an equal weight on each side of the arch it is stable.

But stone is heavy, and eventually it was replaced by steel. A steel arch is anchored by two stone piers at either end. Resting on the arch are steel columns which hold up the flat deck of a road or railway.

early stone bridge

stone arch bridge

steel arch bridge

Rio Niteroi Bridge

Designing large-scale bridges demands a high degree of skill. When the Brazilian government decided to build a bridge across the River Niteroi, it engaged some British engineers to make the plan succeed. Many major structures of this kind all over the world are built by similar multi-national teams of designers and workers. There are, of course, extra problems. Working in a foreign country is never easy, particularly with engineers and workmen speaking different languages.

The plan

More than nine km (five mi) of the bridge deck had to be built in less than one and a half years. The engineers knew that normal methods of building would not finish the job in time. Using a very simple idea on a huge scale they decided to use a type of construction that had not been tried before on so large a bridge. The center section of the bridge was to be a gigantic box girder.

All the steel sections to support the road deck were made in small units in Britain. A German ship was hired to carry them to Brazil, where they were welded together to form larger parts of the structure. Then came a real stroke of ingenuity. One section was shaped like a boat so that it could carry all the other parts upriver to the growing bridge.

Huge concrete piers, or caissons★, had already been sunk into the riverbed. Now each long section of the steel box girder had to be floated out to its correct position and put on top of the piers. The idea was to lift them into place with special jacks, but what should have been a fairly simple job turned out to be very difficult. The weather had been growing steadily worse in the mouth of the river and work stopped for several days.

Time was now short, and the pressure was really on. For a time it looked impossible, but the workmen managed to meet the schedule.

Box girder strength

Box girder bridges get their name from the beams used to support the roadway. Box girders are light and strong and can be used to make large spans. A box girder design for the Rio Niteroi bridge was chosen because of the nearby airport. The bridge couldn't have high towers because of low-flying aircraft in the area.

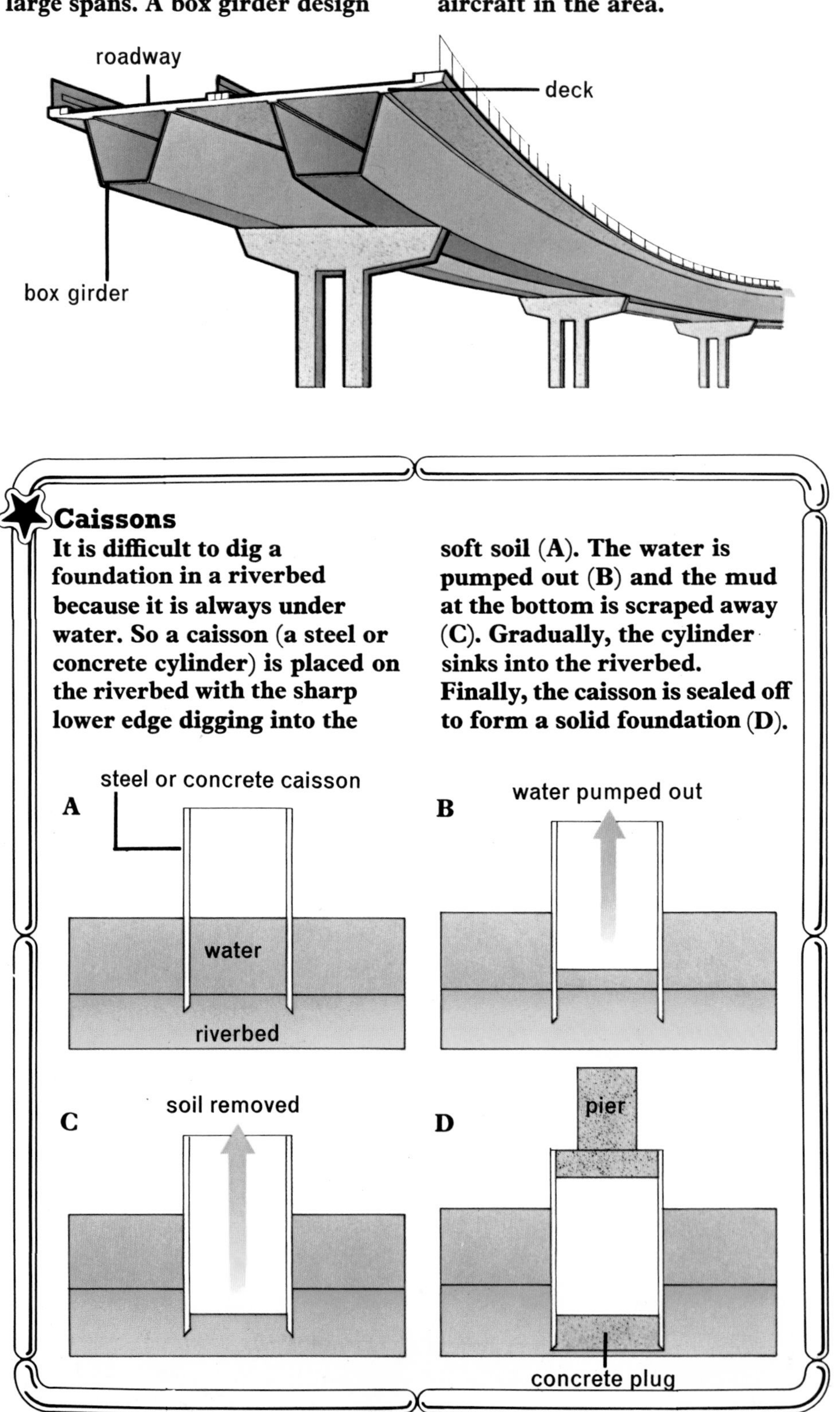

★ Caissons

It is difficult to dig a foundation in a riverbed because it is always under water. So a caisson (a steel or concrete cylinder) is placed on the riverbed with the sharp lower edge digging into the soft soil (A). The water is pumped out (B) and the mud at the bottom is scraped away (C). Gradually, the cylinder sinks into the riverbed. Finally, the caisson is sealed off to form a solid foundation (D).

Using part of the box girder as a barge, one of the other sections is floated out towards the piers in the middle of the river.

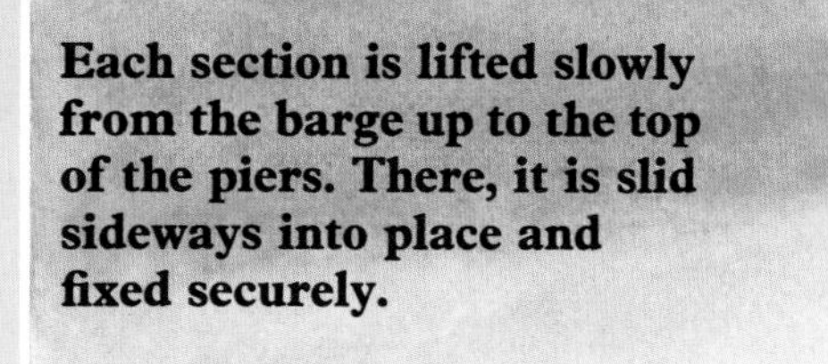

Each section is lifted slowly from the barge up to the top of the piers. There, it is slid sideways into place and fixed securely.

Wide spans were needed near the center of the bridge to give ships enough room to pass underneath in safety. Nearer the shore, shorter spans were built using concrete box beams. All the piers rest on caissons sunk into the riverbed.

Roads

It would be hard to imagine life today without roads—no cars to take us on vacations, no buses to get to school or work, no trucks to carry goods around the country. We all rely on roads most of the time but do not often give much thought to the men who design and build them.

Highway network

Most new roads built today are fast freeways which run for miles without sharp corners or hold-ups like traffic lights. These major roads allow cars and trucks to travel much more quickly than on conventional roads. Planning and building a highway is a long and tricky job. There are rivers to cross, hills to climb, and always the road has to be smooth and fairly straight. To avoid having steep climbs and sudden drops, the engineers cut through the tops of hills and then use the rock to fill in the valleys that have to be crossed.

The earth-movers

All construction work starts with clearing the ground. Building a road means that millions of tons of earth have to be shifted. As the earth-moving machines set to work, surveyors constantly check that just the right amount of soil is taken away. The planners have already decided how many bridges and tunnels they will need, and work begins on them well in advance.

When the shape of the road has been carved out of the countryside, the road base material is spread out evenly by graders. The road surface is laid on top of this. Only two types of surface will stand up to the pounding of heavy trucks traveling at high speeds. One is concrete, the other *blacktop*, often called tarmacadam or asphalt. Concrete lasts longer, but to get good drainage the surface has to be 'grooved', which makes for a noisy ride.

Bridge
Bridges are usually the cheapest way of crossing water.

Cutting
Expressways cut through hills to avoid steep climbs.

Embankment
Soil from the cutting is often used to form an embankment which carries the road over valleys and low ground.

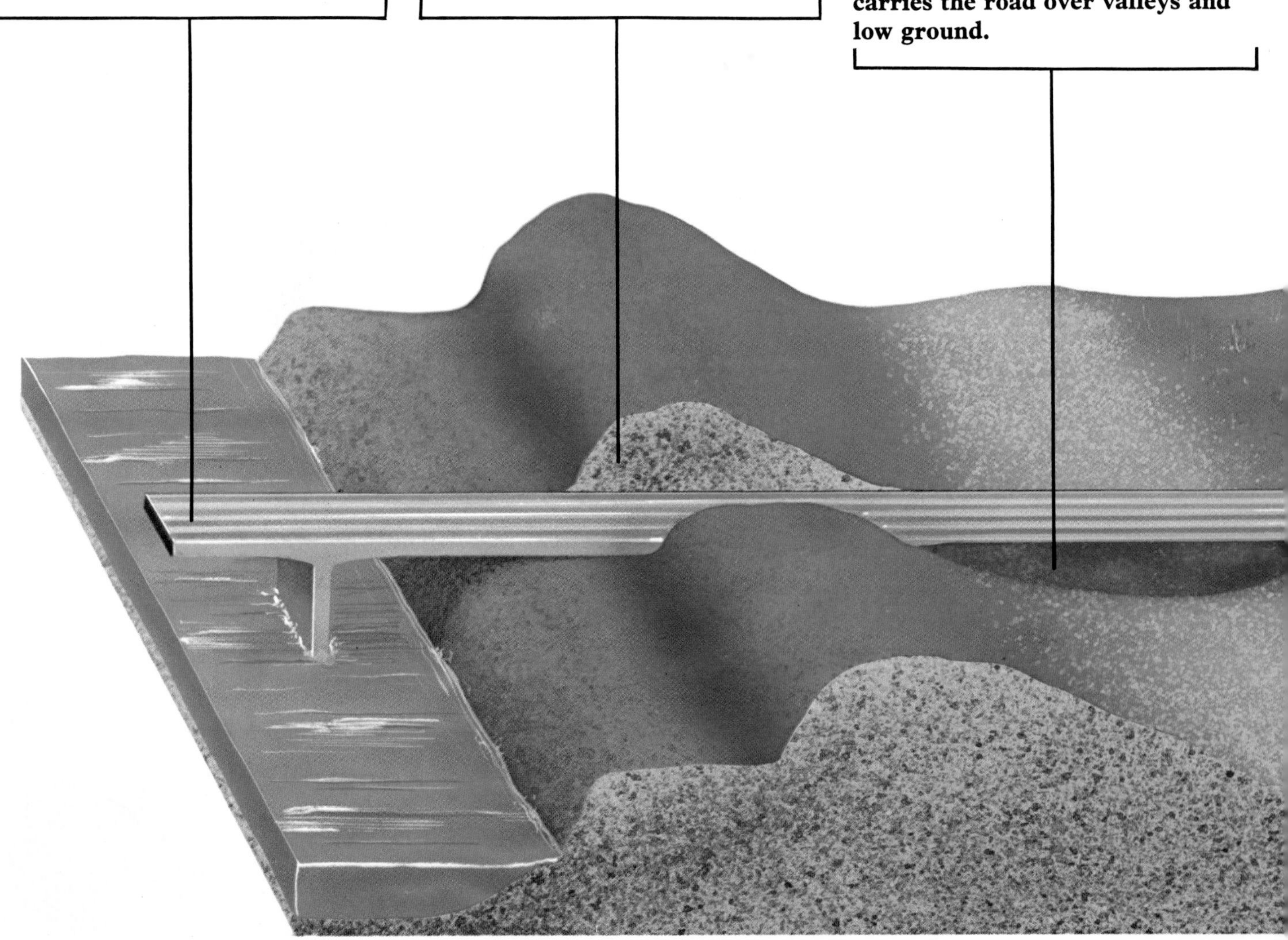

Carving a road through the Amazon jungle calls for great engineering skill.

When highways meet, an expensive and complicated junction often has to be built.

Traffic analysis

Road planners sometimes spend years analyzing the need for a new road. Here, you can see one way that planners use their findings. Every morning 500 cars from a suburban housing development pass through a village en route to a factory. With a new road going directly from suburb to factory, less traffic will congest the village, and streets and traveling will be quicker and safer.

Viaduct

If the ground is soft or marshy, a bridge is built. A bridge over land is called a viaduct.

Tunnel

If a hill is too high to be cut through, a tunnel is built.

Cloverleaf

This kind of junction allows cars to get from one highway to another without stopping. But they are expensive to build and use up a lot of room.

Tunnels

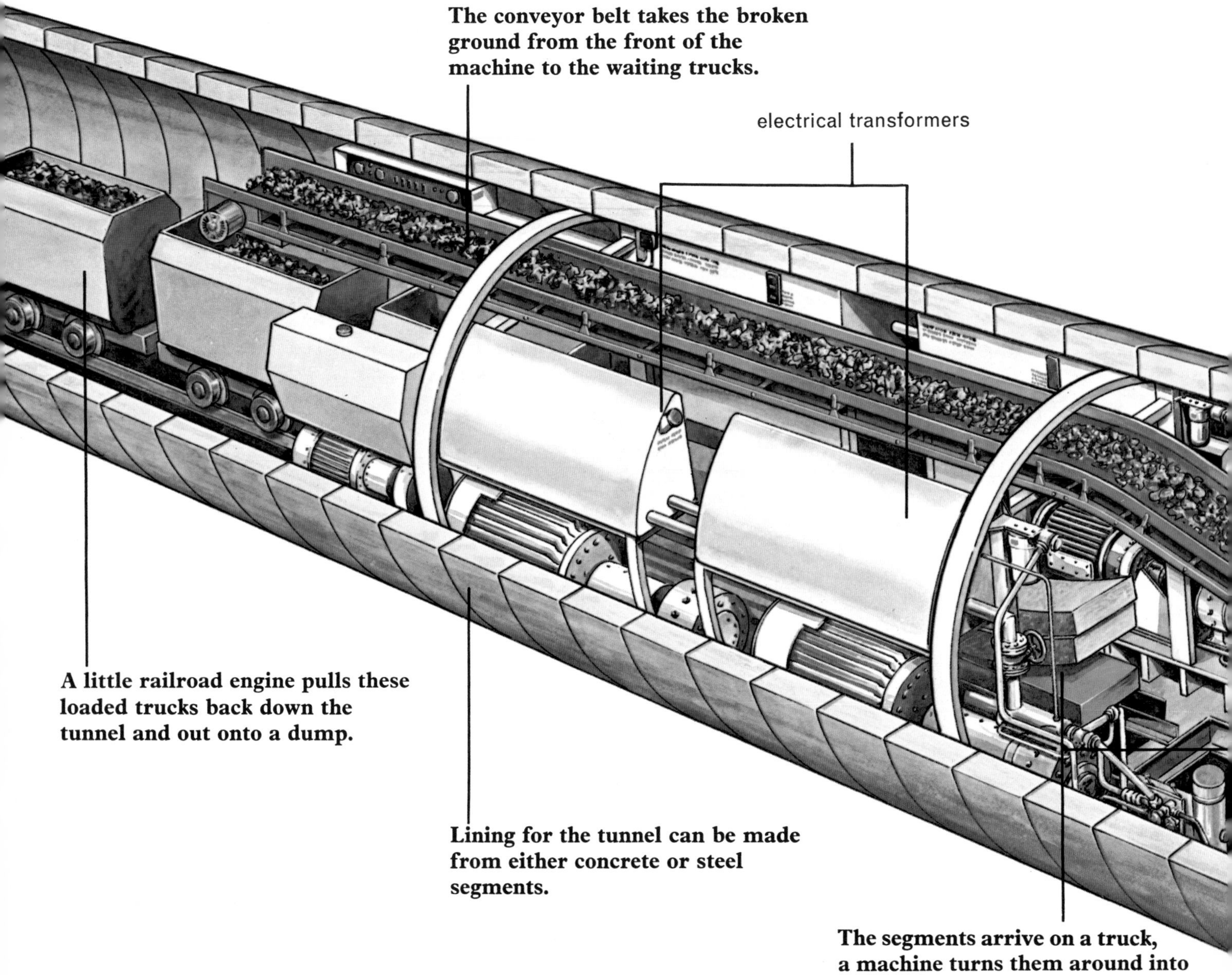

The conveyor belt takes the broken ground from the front of the machine to the waiting trucks.

electrical transformers

A little railroad engine pulls these loaded trucks back down the tunnel and out onto a dump.

Lining for the tunnel can be made from either concrete or steel segments.

The segments arrive on a truck, a machine turns them around into place, and they are bolted together.

There are all kinds of tunnels. They ferry cars, trucks, trains and even boats through mountains, under cities and beneath rivers. Tunnels also carry water, gas and electricity right beneath our feet in towns and in the countryside.

Methods of tunneling

Not all tunnels are built the same way, but tunnels built through weak or soft ground have a lining of steel or concrete to keep them from collapsing. To make the tunnel, men work in a *shield* which is a tube of steel or wood open at both ends. As it is pushed along, the men simply gouge out the soil. The soil is passed back in trucks on a little railroad. As soon as the shield moves forward, the lining is put in place behind it. This method is often used for tunneling beneath cities and rivers. A much messier but simpler way of building a tunnel is to dig a deep trench. A tube is put into this trench and then covered over. This method is not often used near towns.

The modern way to tunnel is to use a *mole* like the one shown above. But even the mole cannot cope with hard rock. The only way to cut a tunnel then is to drill holes into the rock face and fill them with a *charge* of high explosives, removing the rubble afterward.

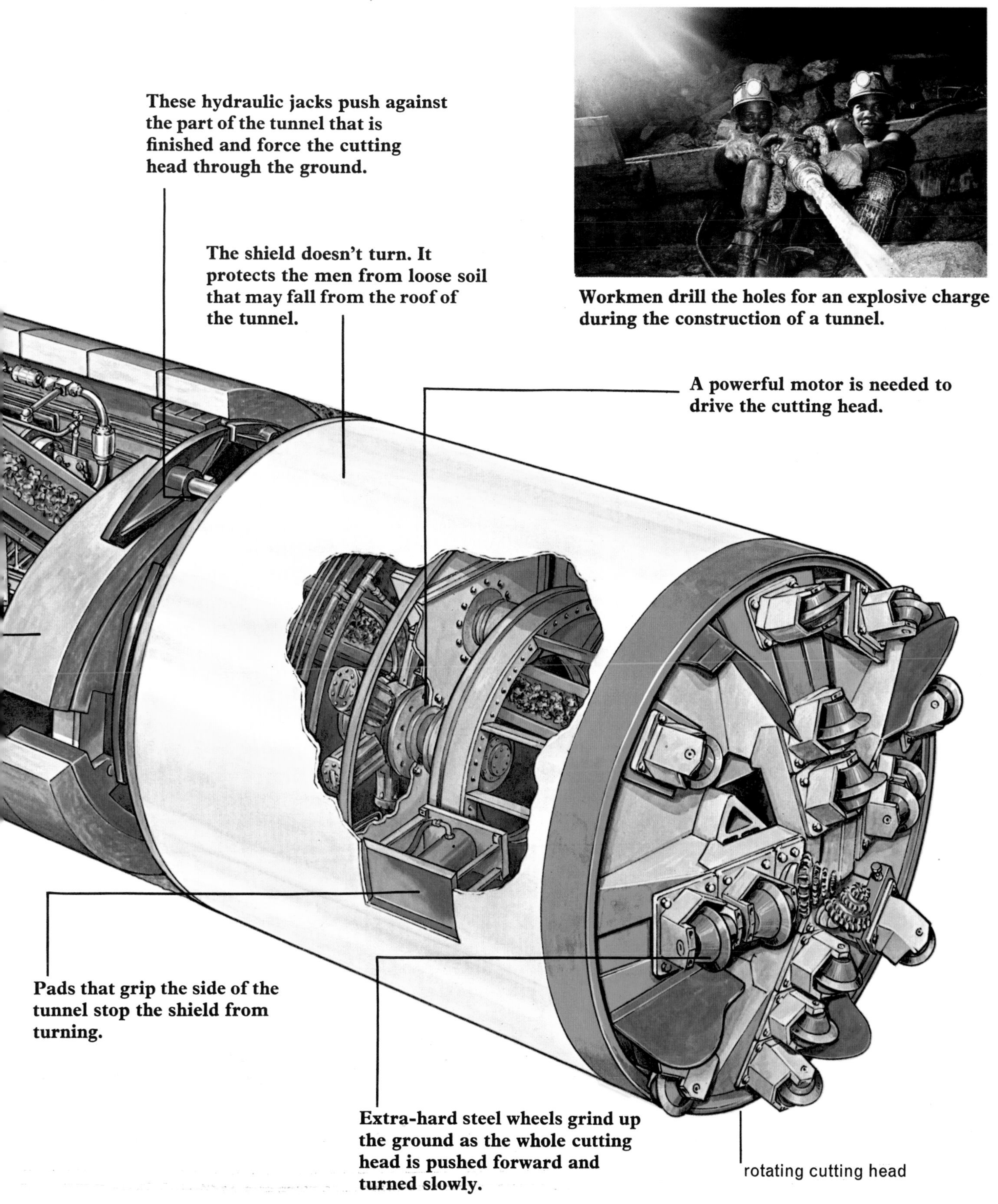

Workmen drill the holes for an explosive charge during the construction of a tunnel.

Hong Kong Harbor Tunnel

For years Hong Kong had to rely on its ferries to connect the two parts of the city. Only recently was a tunnel built under the water that gives fast and reliable access for the first time. No more waiting; just drive through the tunnel and you're across in minutes.

Fast and inexpensive

The tunnel consists of a double-steel tube made up of units joined together in a long chain. Each unit was built on dry land, then floated out into the harbor and sunk ★ into the tunnel trench. On top of the sunken sections went a layer of concrete which was heavy enough to keep the tube from rising to the surface again. Now divers went down to break open the seal leading from the newly laid unit into the section next to it. Both ends of the tunnel were joined to the land by short sections of concrete, then the tunnel was ready for use.

It is easy to see that this tunnel was fairly simple to build. It didn't cost very much either! Ships are in no danger of hitting it as they pass over, because the water is deep.

Engineers inspect one of the twin steel tunnel units before it is floated out into the harbor. The tube is used for pouring concrete.

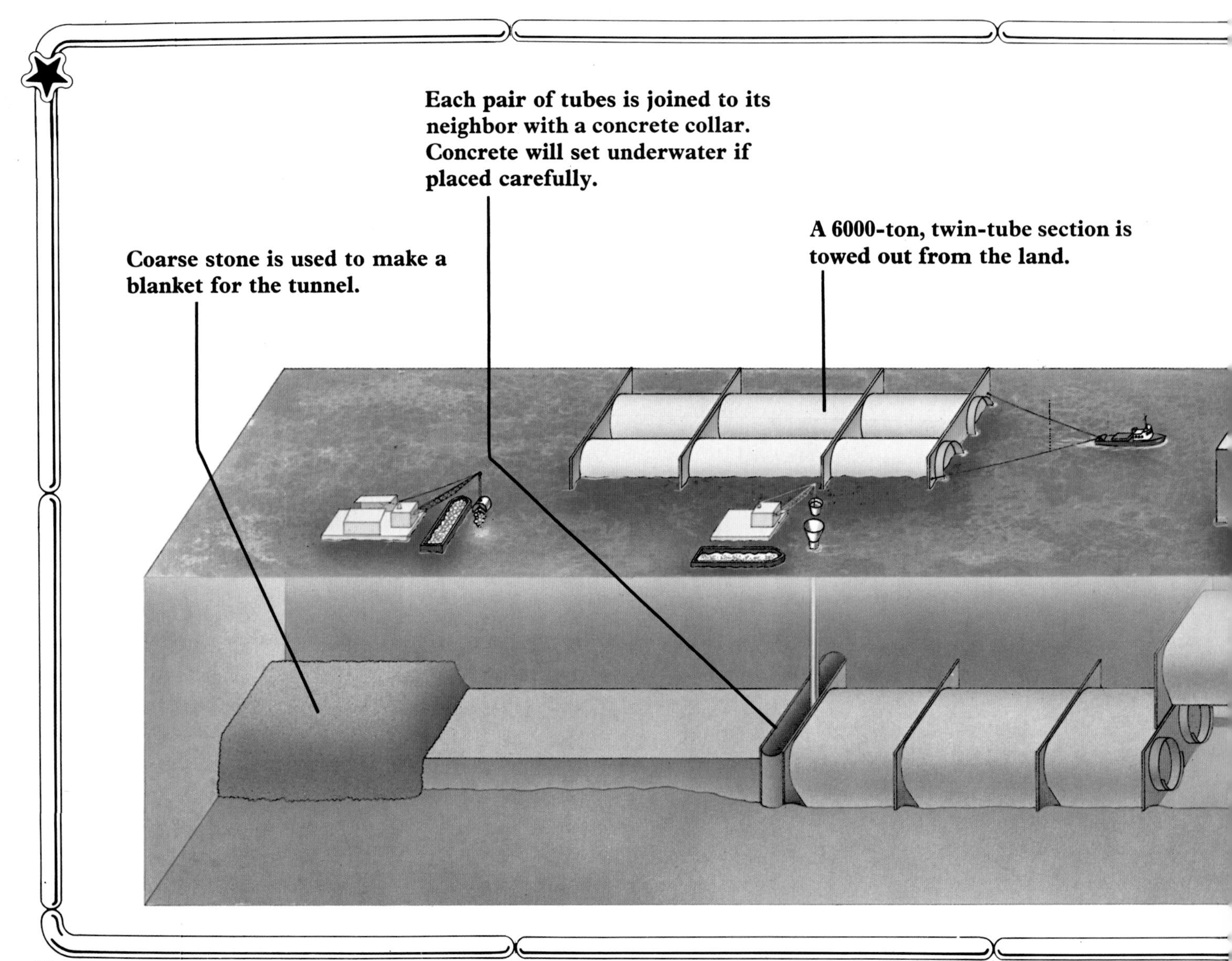

Inside the tunnel

The twin tubes of steel sit in their jacket of concrete on the riverbed. A blanket of stones and gravel keeps the tunnel safe from passing ships. Along each tube runs a concrete road.

Concrete is added to the top of the steel tube unit to make it sink into place. It is lowered gently by the lay barge.

A barge lays a weak concrete base in the trench to give a flat surface to work on.

A dredger digs the trench.

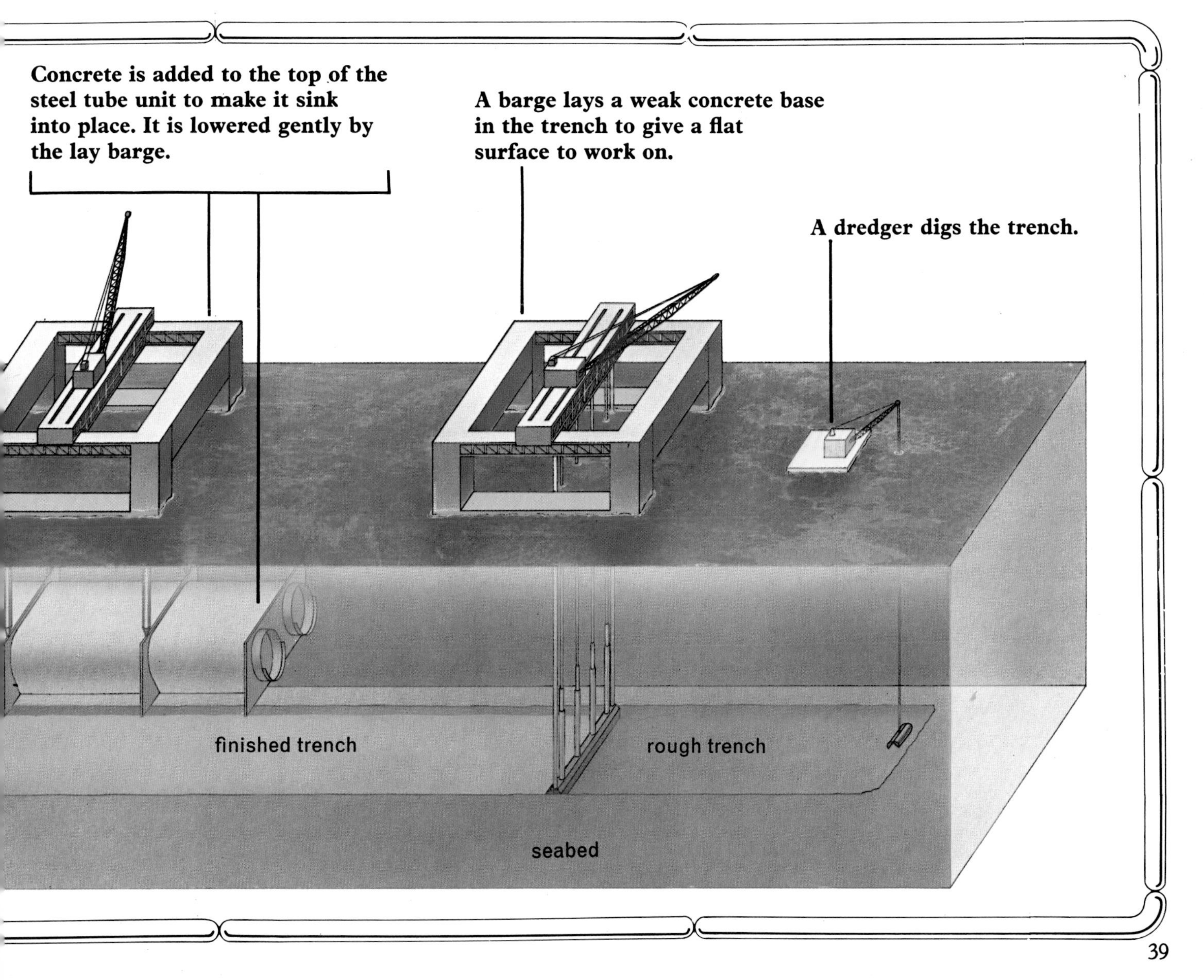

Tarbela Dam

In countries where the climate is hot and dry, rivers are vitally important for watering crops. But how to keep them from drying up in long, hot seasons, or flooding over after heavy rains? Obviously, the answer is to control the amount of water. That's one reason why dams are built.

A dam across the Indus

In northern Pakistan, engineers battled to erect the huge Tarbela Dam across the mighty River Indus. The Tarbela used the simplest dam-building method of all. First the Indus was diverted from its course into a specially made channel. Then work could begin on shifting the enormous amounts of earth needed to dam the water. Slowly the dam rose in height, although progress was often hampered by devastating spring floods, torrential summer rains and freezing winter nights. Many months later the diversion channel was closed and the waters of the Indus were allowed to build up behind the embankment.

Threat of disaster

As the water built up in the lake, parts of the dam began to leak. Boats were sent out rapidly to plug the holes with thousands of tons of stones. When the engineers next tried out the irrigation and power tunnels which had been built into the base of the dam, the force of the water ripped house-sized pieces of concrete from the tunnel walls. This time, all the water had to be drained from the dam to repair the damage. But today, the problems are over.

Make your own dam

You can build your own rock-filled dam in a stream. Take some large stones and make a pile of them between two low banks. Use smaller stones on the upstream face. The water will rise behind the dam but it will leak. To seal it, use sand or soil. Drop it into the water just behind the dam on the upstream side. The small particles will get sucked through the dam and will block the holes. You can do the same at the seaside if you find flowing water.

Four tunnels, 800 m (half a mile) long, carry water for irrigation and for the power station.

At a height of 148 m (485 ft) the main dam is one and a half times as tall as the Statue of Liberty.

This part of the core of the dam is made of fine material which stops water from seeping through the dam.

The dam supplies water to farmers in the surrounding countryside.

The lake created by the dam is 80 km (50 mi) long.

This cutaway shows how the fine material continues under the reservoir to stop water seeping under the dam.

Spillways let excess water escape without going over the top of the main bank.

The main dam is 2.7 km (nearly 2 mi) long.

This high ground has been used as part of the dam.

An extra tunnel for use when the water level in the reservoir is low.

Grand Coulee Dam

One of the largest concrete dams in the world is the famous Grand Coulee on the Columbia River in the state of Washington, completed in 1942. It is a concrete gravity dam. Instead of an embankment of soil and rock, a strong concrete wall is built. The wall stays in place because of the weight of the materials used in its construction.

Like many dams, the Grand Coulee was built to irrigate farmland. The water is now also used to generate electricity.

Under pressure

Plans were drawn up to cut off part of the river valley with a temporary dam. The river was then forced to flow around it. Behind the dam, workmen dug down to rock that was solid enough to make a foundation. Then they extended the temporary dam farther out into the river, but disaster struck. The temporary dam was smashed open by the great pressure of water. Day and night men worked to fill in gaping holes, and eventually they won the battle. It was by no means the final battle they had to fight before the last of the 22 million tons of concrete for the main dam had been added.

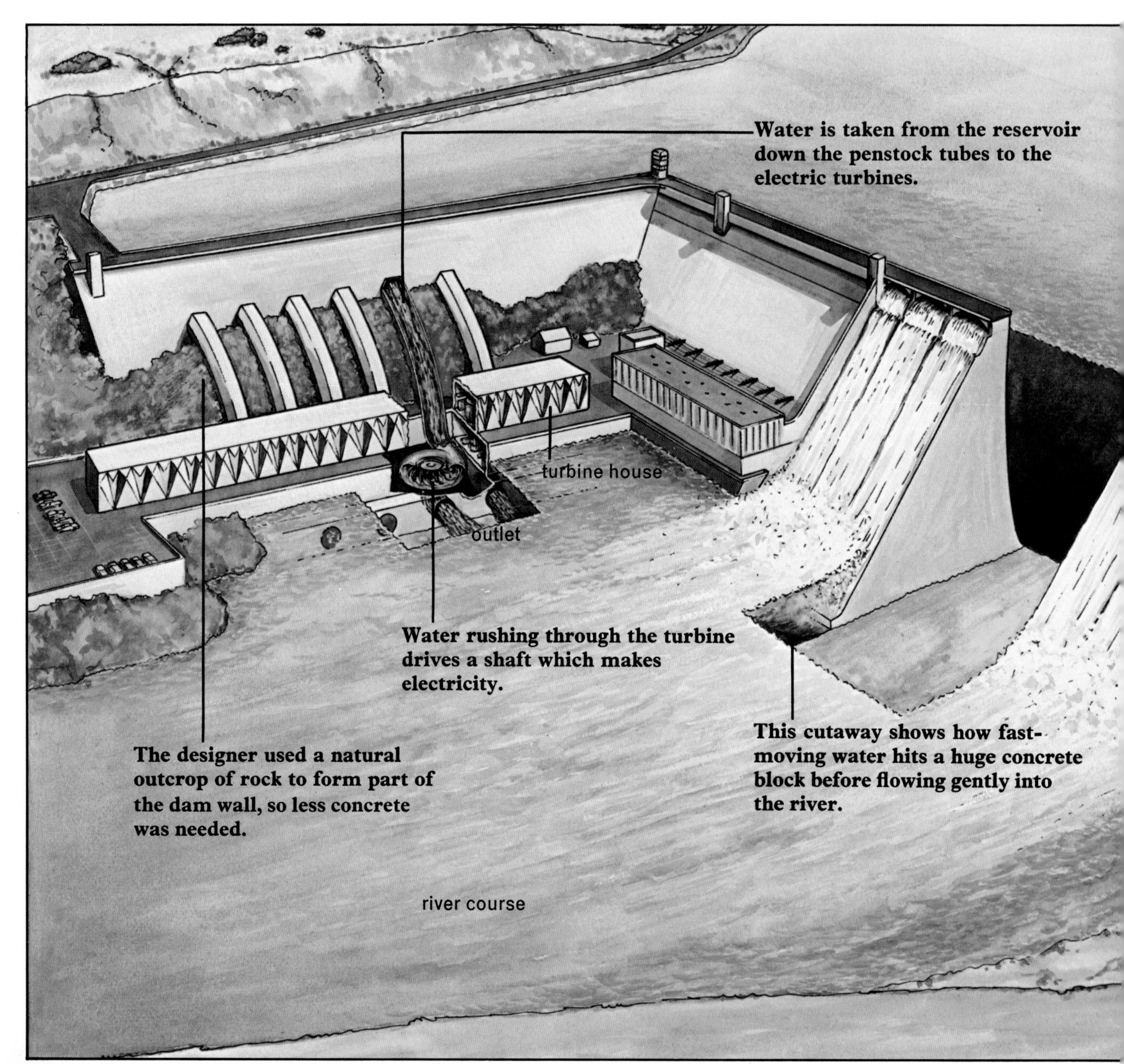

Which dam ?
To build a wall of rock of the same height and strength as a concrete dam you would have to use far more material. The gently sloping sides of a rock bank cannot be made any steeper without their sliding like an avalanche.

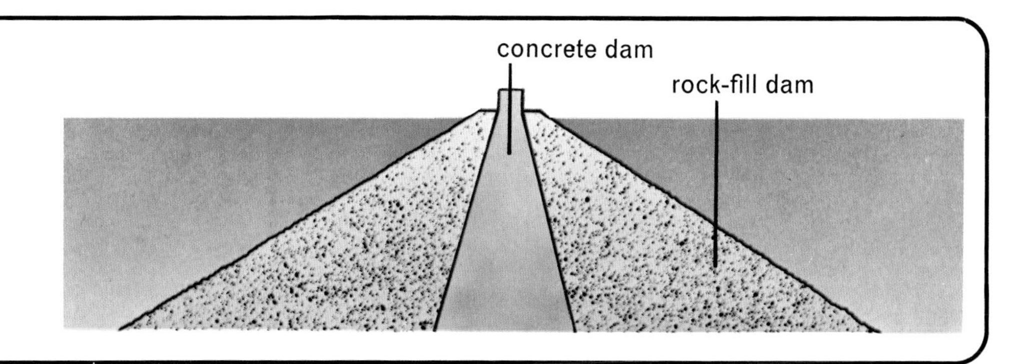

Excess water pours over the spillway and flows on downstream.

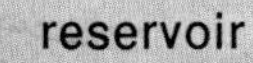

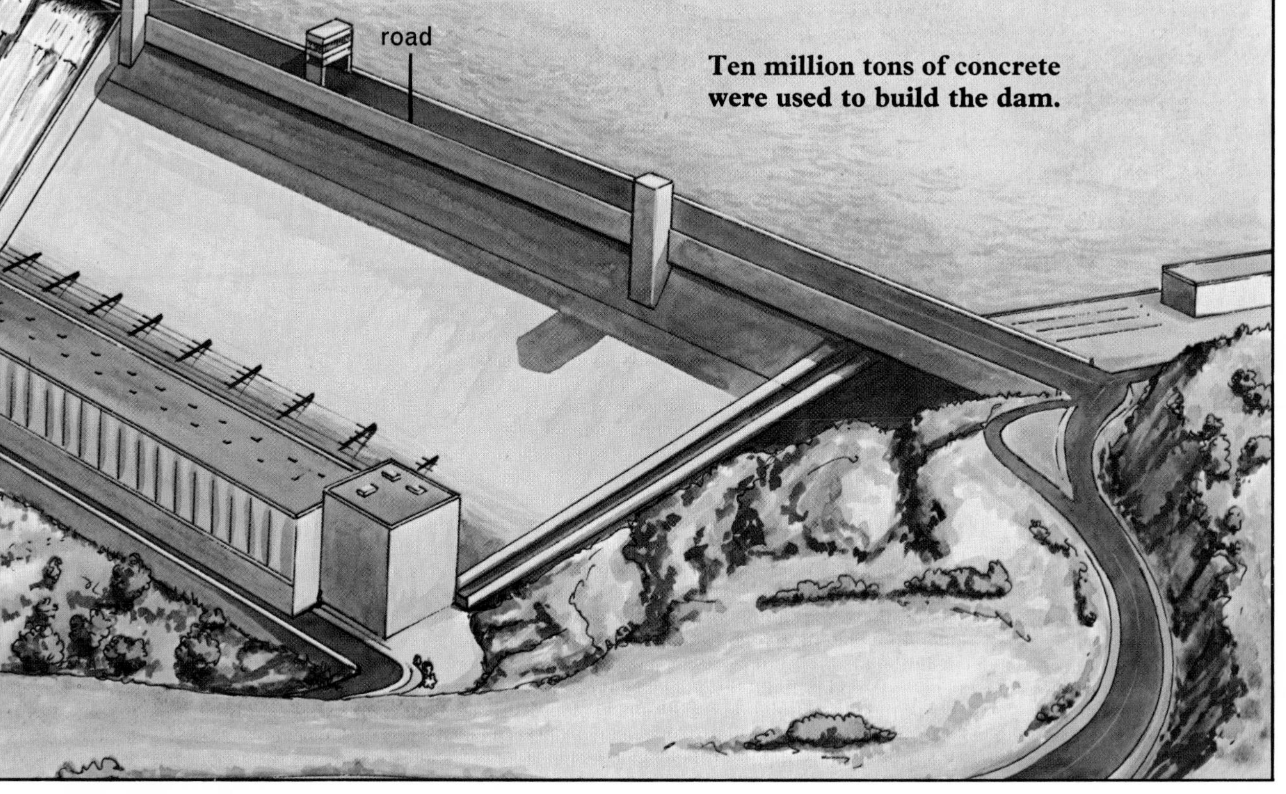

Ten million tons of concrete were used to build the dam.

Karun Dam

Dam building is costly and time-consuming. But, given the right conditions, there is a less expensive dam design than the ones looked at so far. It is called an arch dam.

How does it work?

The best location for an arch dam is a deep, steep-sided valley. The arch dam relies on its arch shape, not on weight, to hold back the vast amount of water in a reservoir. It works just like an arch holding up a bridge, except that the arch of the dam lies on its side. The sides of the valley containing the dam hold it firmly in place.

An unusual diversion

One ideal site for an arch dam lies high up in the mountains of Iran. It is on this site that the Karun Dam is located. As with any dam, the first job was to divert the river. At the mountain site there was little room in the bottom of the river valley, so the engineers decided to build a tunnel to carry the river underneath the dam site while construction was going on.

With the river out of the way, work could begin on the main dam. Two huge grooves were first cut into the sides of the valley and filled with concrete. Engineers call these structures *thrust blocks* and they are built to hold the dam in place. Next, starting at the bottom of the valley, the 185 m (600 ft) wall of concrete was built up. Getting the shape absolutely right was critical. The wall had to curve up the valley toward the flow of the water *and* lean outward at the top. Inside the wall were tubes called penstocks for carrying water to the electricity generators.

A pleasant surprise

When the dam was complete, and all the costs had been added up, it became clear that the dam had cost less than first thought. This cannot be said of many projects today!

Dam builders often have to work in isolated parts of the world, far away from towns. They have to withstand the scorching heat of the sun by day and bitter frost during the night.

Air vents help the flow of water in the penstocks.

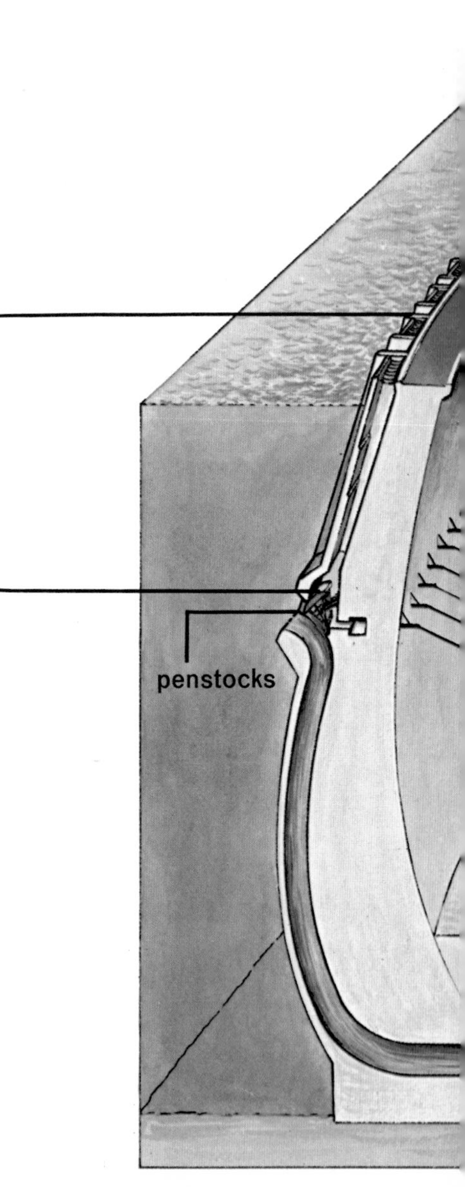

Steel gates control the flow into the huge pipes which carry the water to the turbines.

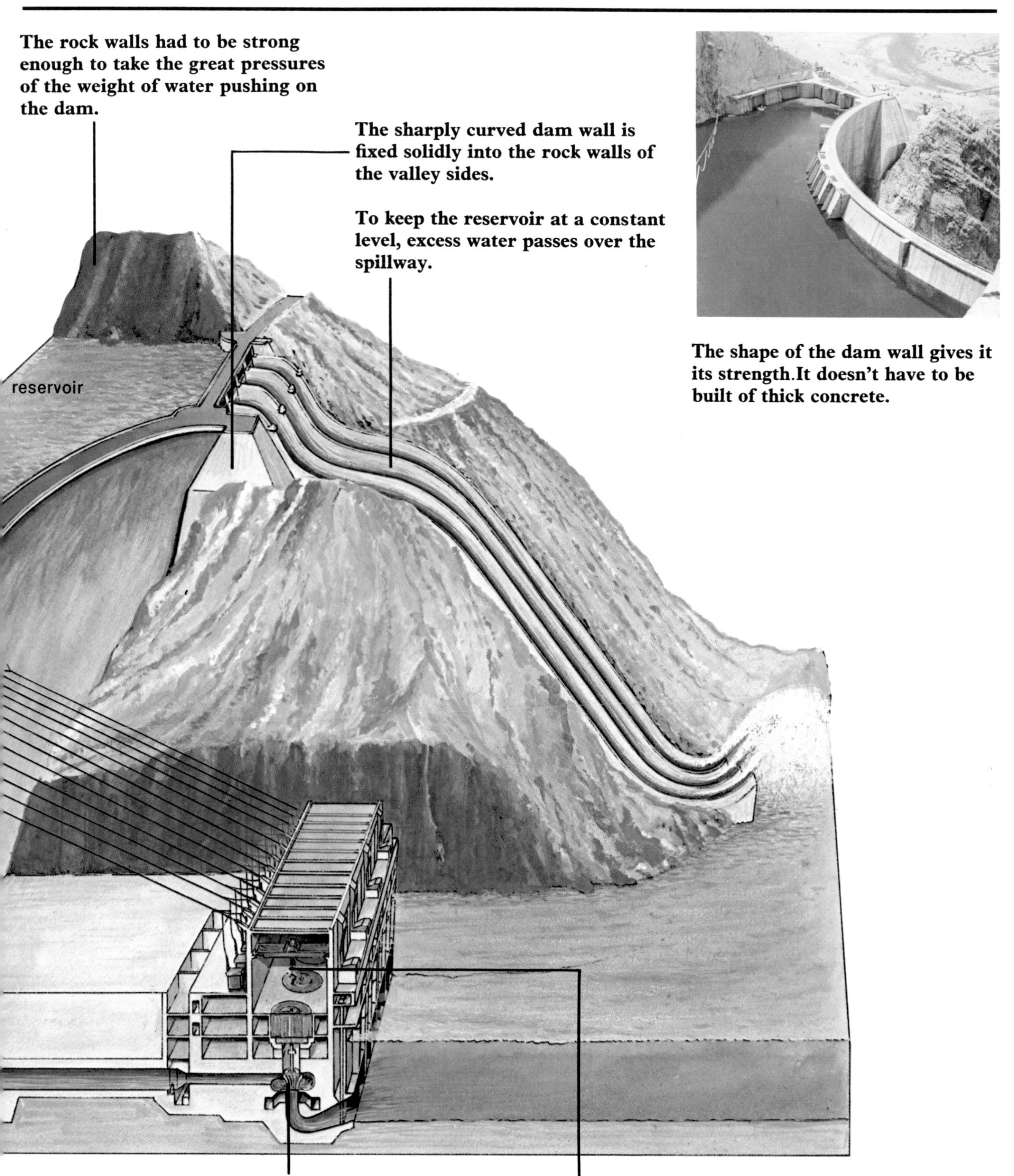

The shape of the dam wall gives it its strength.It doesn't have to be built of thick concrete.

Panama Canal

For centuries canals have been used to transport all kinds of goods. In both the United States and Europe a great network of canals carries very dense traffic right into the heart of industrial areas. Heavy loads can be transported very cheaply in this way and, at the same time, sending goods by canal leaves the roads less crowded.

Canals are really man-made rivers, except that the water is at a constant depth and there are no waterfalls. The largest canals in the world are deep enough to take huge ships. One of the most important ship canals is the Panama Canal. It provides a water highway through the narrow neck of land between North and South America. Ships using it are saved a long and arduous journey around Cape Horn to get from the Atlantic to the Pacific Ocean.

A great engineering feat

Excavations for the canal began in 1881, but by 1900 only about one tenth of the work was done. Thousands of men had to labor for years more, digging away millions of tons of earth. Many of them died from malaria, yellow fever and other diseases then common in tropical climates. The Gaillard Cut (formerly Culebra), an eight-mile channel, alone took 32 years to com-

Two small electric trains pull the large ships through the locks so that the water is not stirred up by the ship's propellers.

Two huge ships pass in the locks along the Panama Canal.

How a lock works

Locks are used to raise or lower ships and boats. Here a small boat is climbing. The lower gate is open and the water inside the lock is at the same level as the lower part of the canal. The boat enters and the gate is closed. A small flap (sluice) in the upper gate is opened to let in water from the upper level of the canal. The boat rises as the water level rises. When the two water levels are the same the upper gate is opened and the boat goes on its way. The lock is now ready for the next boat.

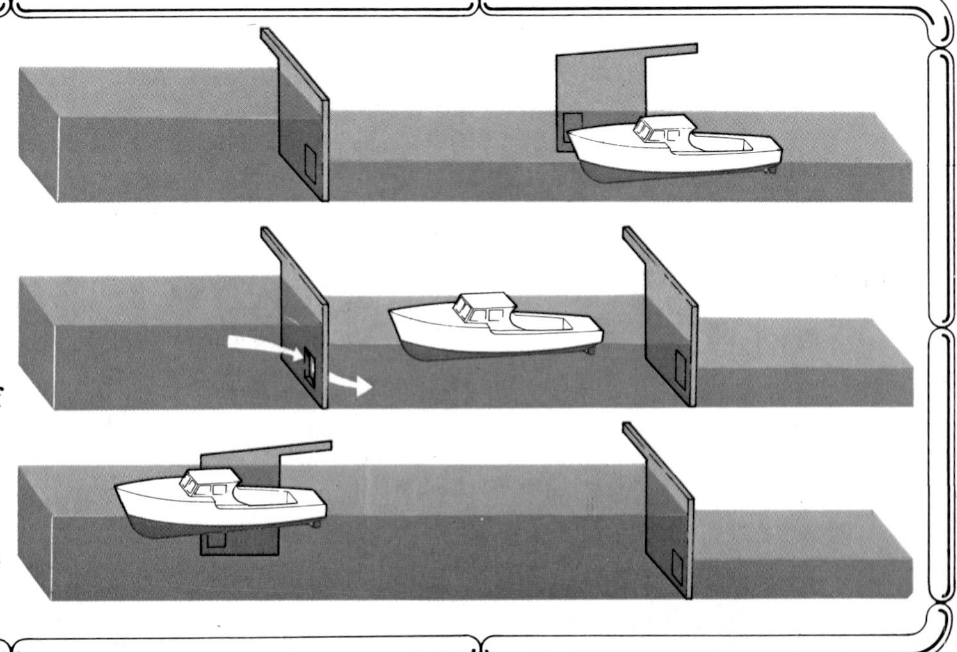

plete. A trench 70 m (227 ft) deep and 185 m (600 ft) wide had to be dug out. In those days the digging machines were powered by steam, and an army of 50 of the noisy puffing monsters had to be used. The soil was taken away in trucks on a specially built railway. Some of the material from the Cut was eventually used to build the two dams that form the lock systems★ at each end of the canal, which opened in 1914.

Bypassing the Cape of Good Hope and Cape Horn cuts thousands of miles off a journey. For this reason, the Suez and Panama canals were built.

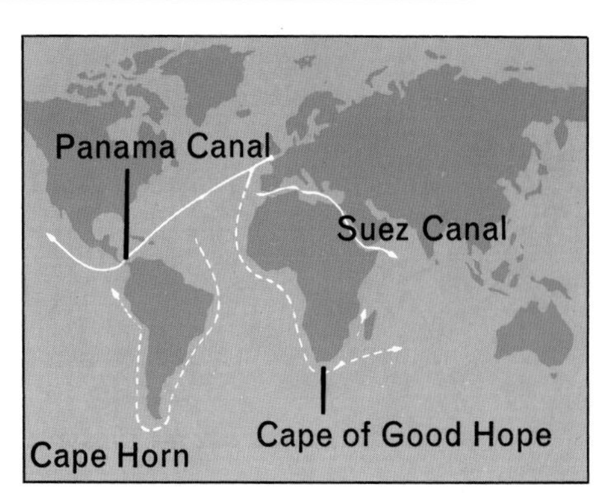

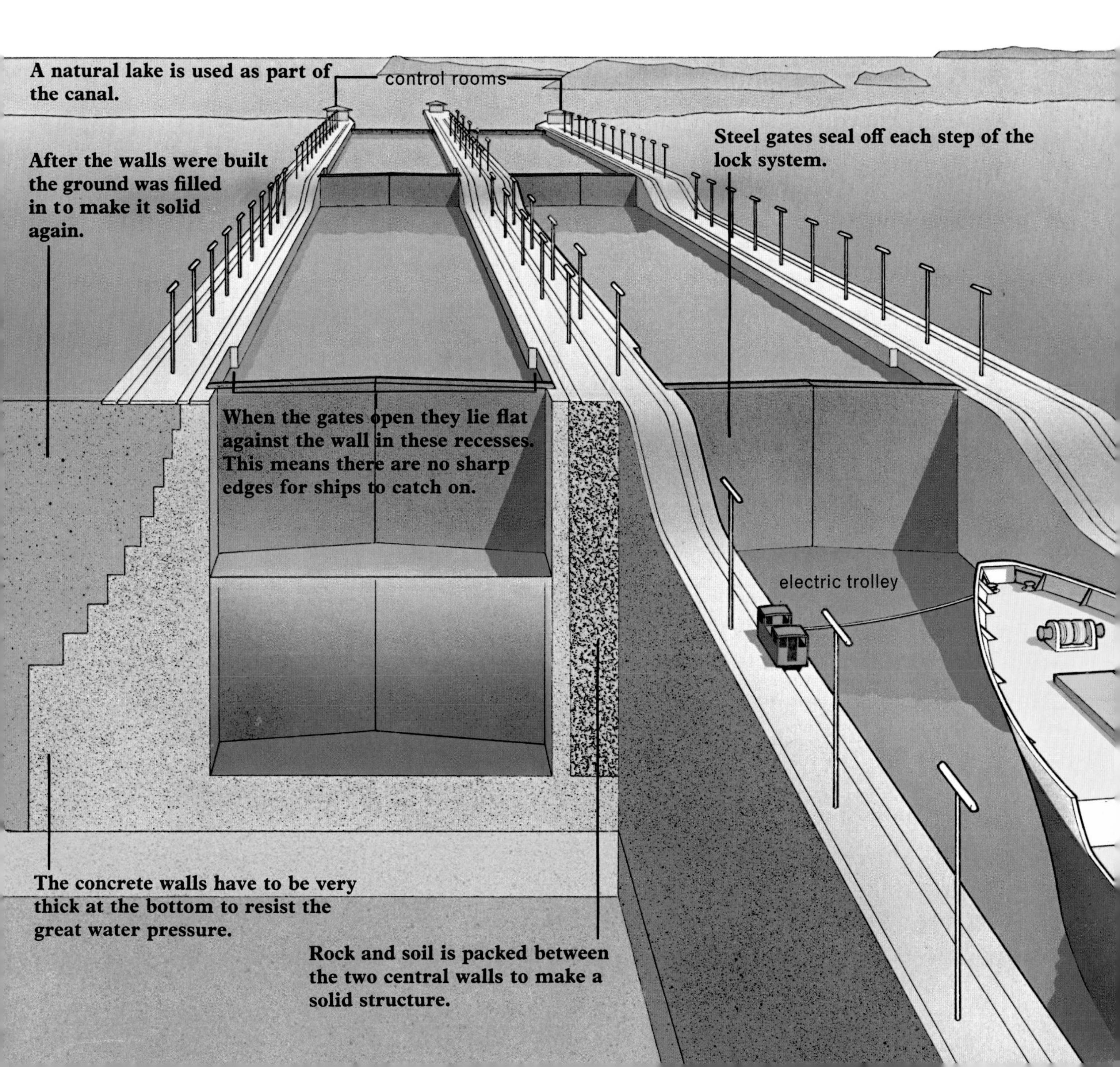

Scharnebeck Ship Lift

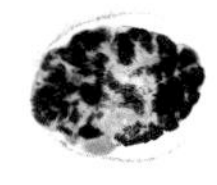

From deep in the heart of Germany's industrial center there runs a new transport route right out to the sea. Ships and barges can now travel nearly 200 km (125 mi) by canal. But where the canal joins the River Elbe for the last few miles of the route there is a big difference in water levels.

Giant bathtub

Mighty locks like those used on the Panama Canal would have solved the problem but the engineers found a new solution. They built a ship lift—a giant elevator that picks up a ship and cargo in a tub of water and carries it 38 m (126 ft) up to the level of the river above. In the bad, sandy soil of the area it was important to build very strong foundations to carry the huge weight of the finished structure. On the lower side of the lift, where the canal finishes, two large steel gates were built to hold back the water. At the top a special basin was built to join the ship lift to the river. There is another steel gate to hold back the water here.

Shortcut

Four huge concrete towers were built to house the lifting equipment. Two large concrete boxes or tubs, big enough to hold a 1500-ton ship, were built between the towers. Each tub is fitted with electric motors and gears which allow it to climb up ladders built into the towers. The result is that, instead of waiting for many minutes for a lock to fill or empty, a ship can get up or down in just three minutes.

The upper level of the ship lift towers above the road below.

Each concrete tub is 100 m (330 ft) long and 12 m (40 ft) wide.

With the gates opened a ship can enter the tub. When the ship is safely inside, the gates are closed. It is now ready to be lifted.

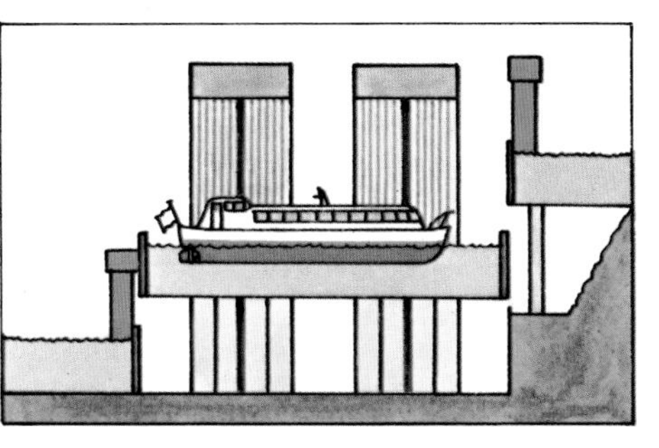

Electric motors raise the tub and the ship together up to the level of the top canal. The tub carries enough water to float the ship.

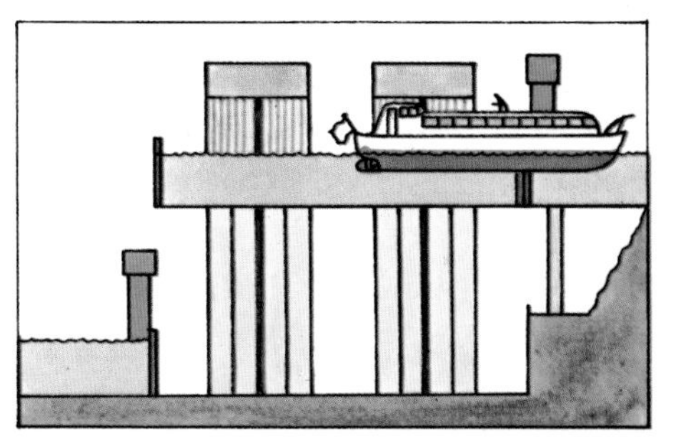

The gates leading to the canal are opened and the ship sails away. A ship lift really does the same job as locks on a canal.

Steel gates swing down to let the the ship pass.

Heavy counterweights balance the tub as it goes up and down.

Another steel gate holds back the water of the upper canal level.

Cooling towers

Probably the most familiar sight at power stations throughout the world is the cooling tower. The job it does is a vital one. It is used to cool the huge quantities of water that circulate through the power generators★.

The shutter

Cooling towers are made of concrete. Their walls are thin because the tower does not contain anything and only has to support its own weight plus the force of any wind. Building starts with columns around the base, on top of which a huge ring of concrete is formed.

A circular form is assembled all the way around this ring and concrete is poured into it. In this way the wall grows taller. When a few feet of the wall have been built, the form is jacked up and more concrete is added to the ring.

The tower becomes narrower the higher it climbs. Pieces of the form are therefore removed to allow for this change in shape. Halfway up, the tower spreads out again, and so pieces are put back in the form. At the top the form is dismantled and taken away to be used on the next job.

The form climbs up casting one complete ring of the tower at a time. Here, a second 'skin' is being added to a finished tower.

Water is heated in huge coal-fired boilers in the building with the single chimney. The stream drives electricity generators. It is then cooled in the eight towers before returning to the boilers.

★ Inside a cooling tower

A power station cooling tower works just like an automobile radiator. In a car, hot water from the engine is fed into a steel compartment called a radiator. A fan sucks in cold air through holes in the radiator. The air becomes hot and the water is cooled.

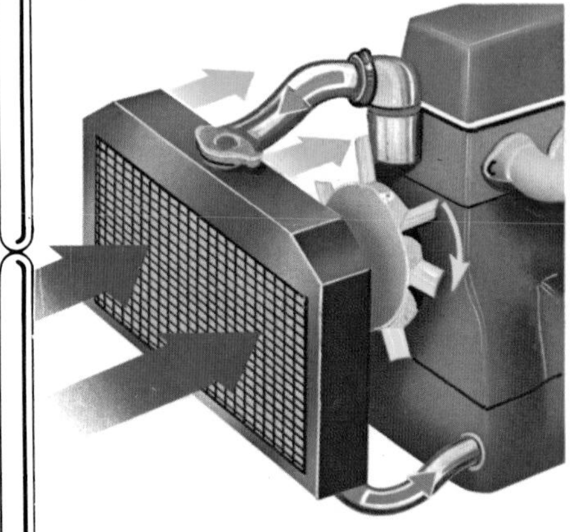

At the power station there is no fan to draw the cold air over the hot water. Hot air rises naturally, so that even without the specially shaped tower there would be a slow movement of the air. The shape is called a *venturi* and it speeds up the air flow in the tower to give better cooling.

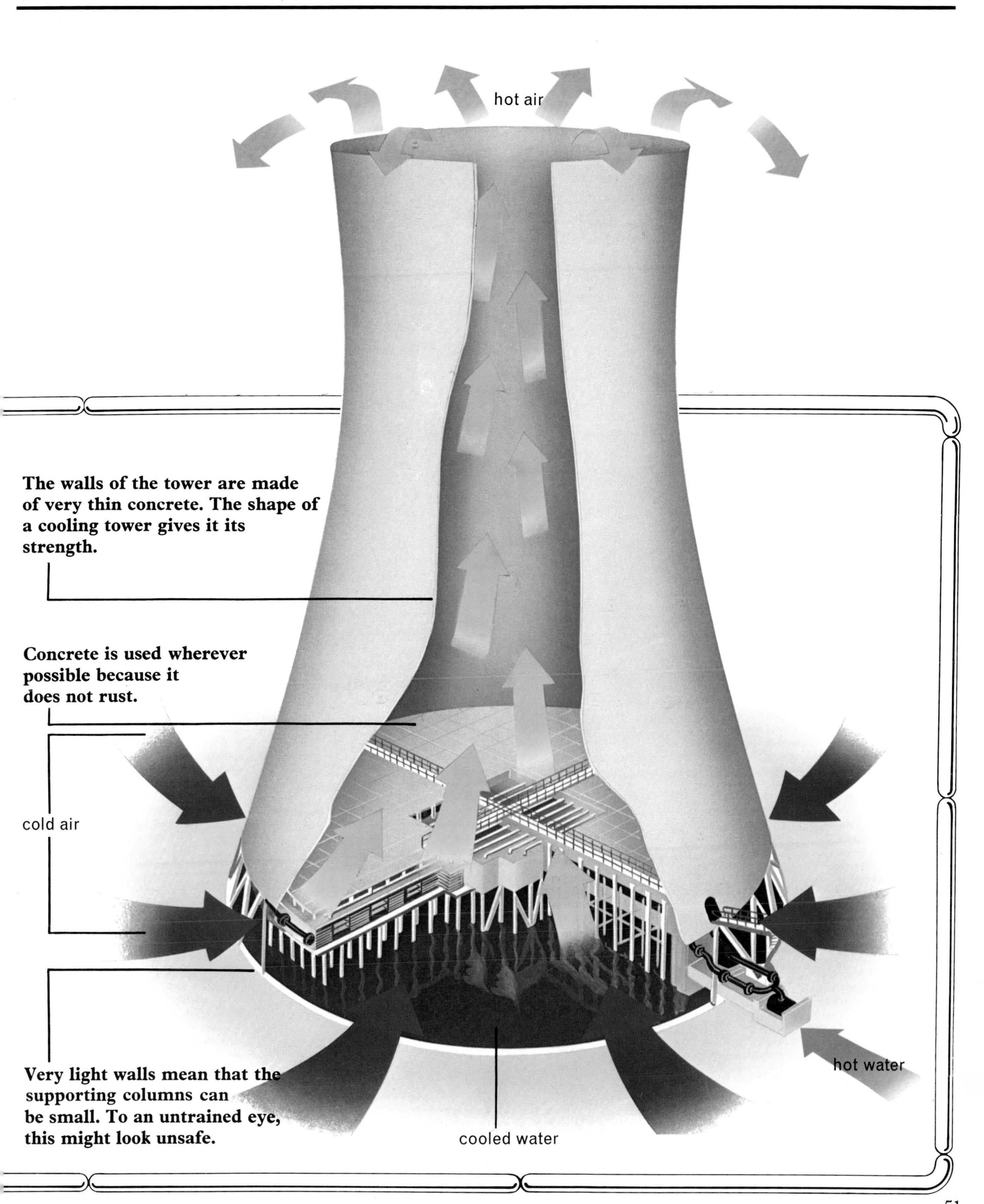
hot air
The walls of the tower are made of very thin concrete. The shape of a cooling tower gives it its strength.
Concrete is used wherever possible because it does not rust.
cold air
Very light walls mean that the supporting columns can be small. To an untrained eye, this might look unsafe.
hot water
cooled water

Ninian Central Platform

You may well have seen photographs of oil platforms in the North Sea. Perhaps the sea looks calm, but ask anyone who works on an oil platform, and he'll tell you how a violent storm can suddenly turn the sea into a surging mass of dangerously high waves. You'd almost think the platform would be engulfed and swept away. It is a tribute to the skill of today's engineers that rigs can operate in all kinds of weather, pumping out millions of liters of precious oil from the seabed.

Dry dock

Ninian Central Platform is a gigantic concrete structure in the North Sea. Its base was built inland in a specially constructed *dry dock.* A dry dock is a small harbor, pumped dry of water, with a gate to keep the sea out. When building is complete, the gate is taken out, the dock floods, and the structure can be towed out by tugs into the sea.

A floating building site

The upper parts of the Ninian platform were added on a little way off the land where the water was deeper. Sometimes storms stopped the work altogether. Once, during a gale, one of the ships that supplied the materials to the floating construction site sank. When the last section had been attached, a fleet of tugboats arrived to pull the giant structure gently out around the top of Scotland and into the North Sea to its final resting place.

Although the concrete base weighs more than 100,000 tons, it floats. Tugs ease it gently out from its dry dock.

In deeper water the concrete walls are built higher and the base sinks slowly under the waves like an iceberg.

With the outer compartments sealed the main core can be cast to its full height.

Powerful tugs pull the floating island to a new site where the water is even deeper.

The dry dock will be flooded when the great concrete gate is taken away. Then the base can be carefully towed out.

The tanks in the base of the platform are flooded. It sinks low enough to get under the deck, which floats on a pair of barges.

After the platform has been towed to its final site in the North Sea, it is weighted down. It now rests firmly on the seabed.

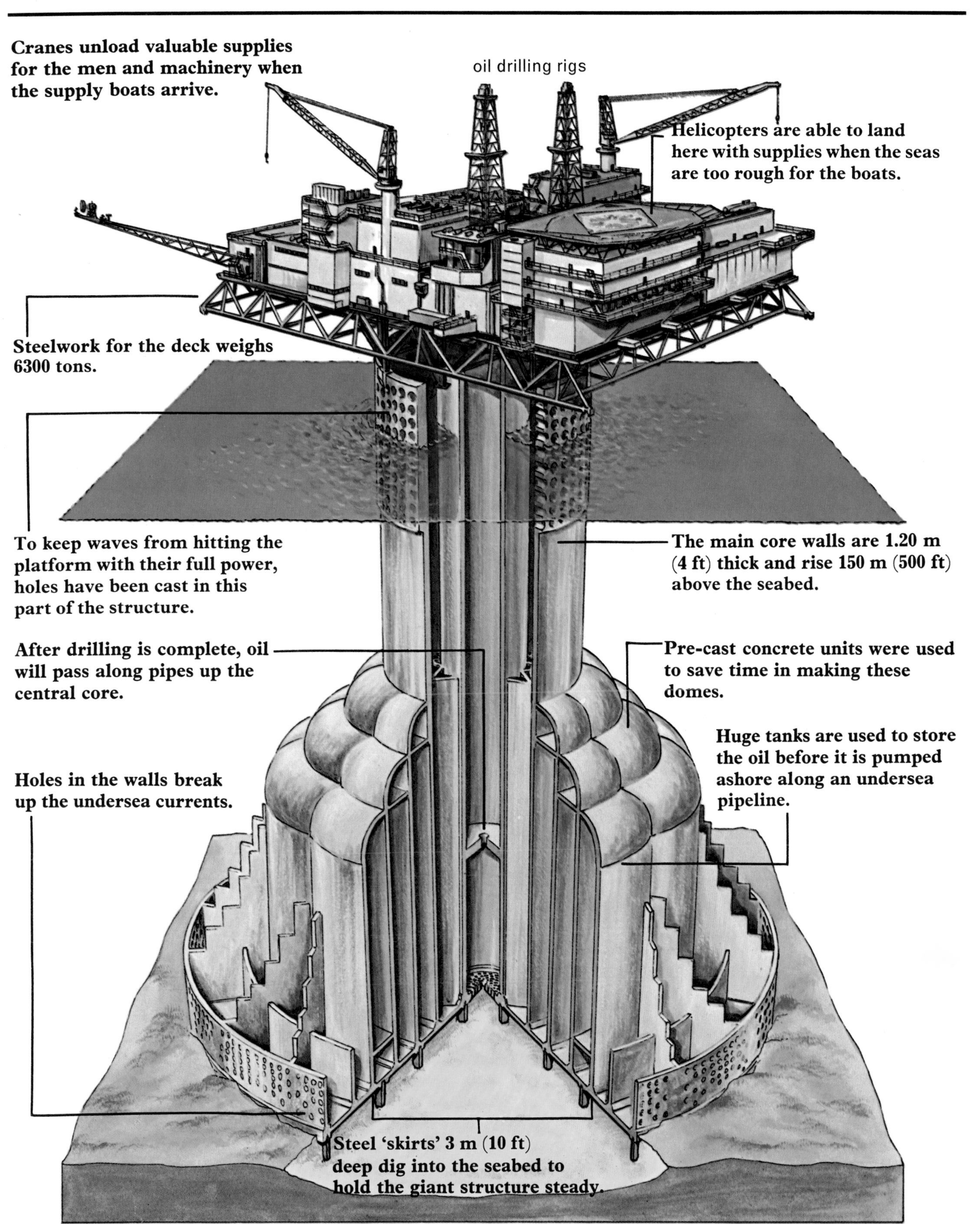
Cranes unload valuable supplies for the men and machinery when the supply boats arrive.
oil drilling rigs
Helicopters are able to land here with supplies when the seas are too rough for the boats.
Steelwork for the deck weighs 6300 tons.
To keep waves from hitting the platform with their full power, holes have been cast in this part of the structure.
The main core walls are 1.20 m (4 ft) thick and rise 150 m (500 ft) above the seabed.
After drilling is complete, oil will pass along pipes up the central core.
Pre-cast concrete units were used to save time in making these domes.
Huge tanks are used to store the oil before it is pumped ashore along an undersea pipeline.
Holes in the walls break up the undersea currents.
Steel 'skirts' 3 m (10 ft) deep dig into the seabed to hold the giant structure steady.

Major disasters

What went wrong?

Imagine the terrible results if a huge dam suddenly breaks up and tons of water cascade down into the valley; or if a giant skyscraper suddenly crashes to the ground. Untold damage may be done to a vast area of land and—worse—thousands of lives may be lost. Fortunately, disasters on such a big scale are very rare indeed. But they have happened.

Preventing disasters

Natural forces like earthquakes, hurricanes, or tornadoes have brought many buildings and structures crashing down. Millions of dollars are spent repairing the damage. But these are certainly rare cases. Usually a collapse happens because of a fault in design or construction. Somewhere along the line the designer's calculations have gone wrong. There may be some kind of force acting on the building which he has failed to take into account. On the other hand, the problem may be that the building team simply has not made the structure strong enough. After every failure, a great deal of trouble is taken to understand what went wrong and to make sure it won't happen again. Builders and designers are learning from their mistakes all the time and making their structures safer.

High winds

These dramatic pictures are from a film taken of the Tacoma Narrows Bridge in Washington as it was destroyed during a gale in 1940. Fortunately, no one was on the bridge at that time. The deck of the bridge was shaped too much like an aircraft wing. As the wind blew, the deck lifted up and down and shook itself to pieces. The disaster made bridge engineers think again about the shape of suspension bridge decks. More recent designs remain stable in high winds. None of the suspension bridges built since the Tacoma disaster has suffered the same fate.

Weak structure

Strong winds were also to blame for the collapse of this cooling tower in Britain. The designers had not made the structure strong enough to resist the buffeting of the high winds.

Pressure build-up
As the water level built up behind the Teton Dam in Idaho in June 1976, the structure suddenly burst. Water had found a weak spot in the dam wall and began trickling through. The trickle grew to a flood which swept away the dam and swamped the valley below. After this failure, engineers inspected many other dams built to the same design.

High-rise disaster
This apartment house in Britain was built like a house of cards. When a gas explosion blasted away one wall near the top of the building, the rest came tumbling down. Now designers make sure that all parts of a structure are tied very strongly together.

Demolition

Have you ever watched demolition workers smashing down the walls of a building? As the crane swings its big ball of steel or concrete, a structure that may have taken years to put up can be demolished in a matter of hours. But why do buildings have to be pulled down?

A set life span

When designing a building, the architect has to make it strong enough to last at least a certain number of years. Prefabricated houses must usually last at least 20 years. Office blocks must be strong enough to still be standing after 30 years, the average house for 65-80 years. Tunnels and bridges must last for about 100 years. The number of years, in each case, is called the *design life* of the building.

Don't think that this means that a tunnel will fall down when it is exactly 100 years old. It means that after 100 years engineers will have to take a close look at the tunnel and decide whether to pull it down or to strengthen it. Very often, little repair is needed. If they decide it does have to come down, that's when the demolition men move in.

In the center of large cities there are a lot of demolition sites. Old offices are pulled down to make way for new ones which will have more room and better working conditions for the people who use them. If the building is well away from others, explosives can be used to bring down the whole building in one go. Occasionally, parts of an old building are preserved for future use in a new one. In that case the demolition work is done, for the most part, by hand, and with great care!

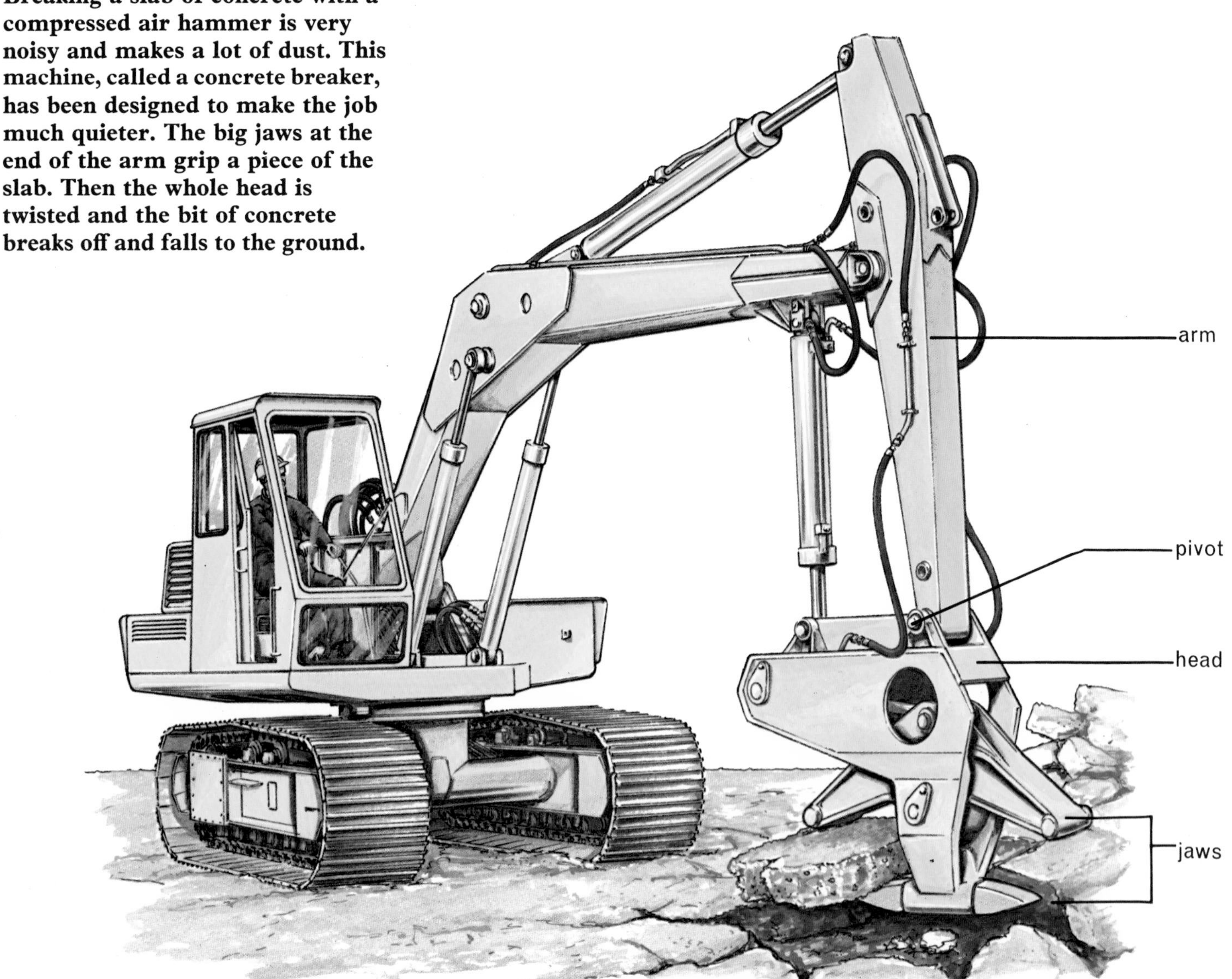

Breaking a slab of concrete with a compressed air hammer is very noisy and makes a lot of dust. This machine, called a concrete breaker, has been designed to make the job much quieter. The big jaws at the end of the arm grip a piece of the slab. Then the whole head is twisted and the bit of concrete breaks off and falls to the ground.

Sometimes, even very new buildings have to be demolished. This skyscraper in Brazil was found to be unsafe and it wasn't considered worth repairing. Demolition men were called in and they decided to use explosives. All 32 stories had to fall within a small area. Structures nearby had to be protected from the blast. Then the first of many very small charges were set. These broke up the columns of the lower floors and the building started to fall. More charges were set off to break up the upper stories as they got near to the ground.

Right: controlled explosions allow demolition men to remove parts of a structure without destroying the whole thing. Only the top of this old bridge in Pennsylvania had to come down.

Building for the future

What will the next century bring? Will the Earth become so overcrowded that people will be living in cities under and on top of the sea? Will there be huge steel structures out in space forming the world's first space colonies? Right now engineers and designers are working toward the future, discovering the technology needed for these ideas.

One way of solving the problem of our overcrowded planet is to build cities out at sea. Engineers have already drawn up plans for man-made islands in shallow parts of the sea. They would provide an ideal site for factories manufacturing dangerous chemicals.

To prevent floods from reaching the heart of great cities, engineers are building huge barriers like this one across the Thames in London. In times of flood, gates swing up into place to form a dam. When the danger has passed, they are lowered again.

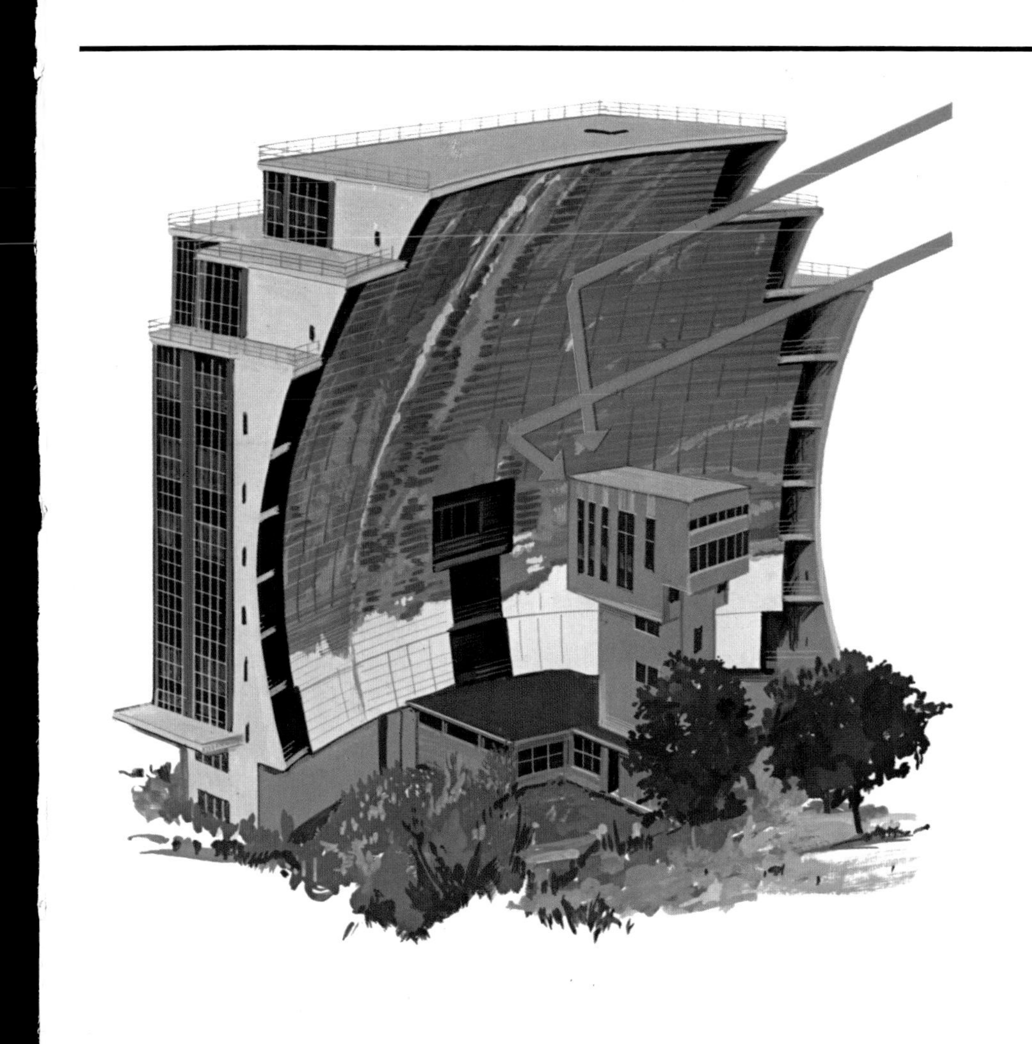

This solar furnace in France looks like a giant mirror. It focuses the sun's rays on the boiler which is housed in the smaller concrete tower. This heats the water, which is then taken to where it is needed. The beauty of this idea is that unlike oil and coal, energy from the sun will never run out.

One day, we shall almost certainly build cities in space and on other planets. The problems of construction may be enormous, but, as has been shown in this book, difficulties don't stop engineers from succeeding.

Contributors

Our thanks go to the following for their help in preparing the illustrations for this book: British Steel Corporation, Central Electricity Generating Board, DEMAG Verdichter und Druckluft-technik, Freeman Fox & Partners, The Greater London Council, Robert L. Priestley Ltd., Westminster Dredging Company.

Photographs

Australian News and Information Bureau 15. Anglo American Corporation of South Africa 37. Bierrum & Partners Ltd. 50. British Institute of Civil Engineers 54. British Steel Corporation 33. Building Research Establishment 55. On loan from the Bureau of Reclamation 43, 55. Canadian National 26. Caterpillar Overseas S.A. 5. Central Electricity Generating Board 10-11, 50, 54. Coles Cranes Ltd. 4. *Construction News* 38-9, 41. Controlled Demolition Inc. 57. Costain Construction Ltd. 10-11. Daily Telegraph Colour Library 34. Fotostudio Cerny 23. Freeman, Fox & Partners 28, 29. Gouvernement du Québec 24. Harza Engineering Company 44. Howard Doris Ltd. 52. John Deere Ltd. 4. J.C.B. Sales Ltd. 4, 5, 9. Behnisch & Partner inside front cover, 20, 21. Ministry of Energy, Teheran 44. National Westminster Bank Ltd. 12. Piano & Rogers Architects 18. Picturepoint Ltd. 34. R.M.C. Group Services Ltd. 5. Robert Harding Associates 46. S. Sperring 48. Tudor Safety Glass Company Ltd. 24. West Virginia Department of Highways 30-1. ZEFA 16.

INDEX